Silkworm Physiology
A Concise Textbook

Dr. Md. Shamsuddin passed his B.Sc. in the year 1967 with Distinction from the University of Burdwan. He passed his M.Sc in the year 1969 with 1st class and was awarded national scholarship during his post-graduate study. He was awarded with Prof. S.N. Mukherjee Memorial medal. He was awarded with Junior, Senior and Post-Doctoral fellowships by the Council of Scientific and Industrial Research,New Delhi for Research work on "Some aspects of differentiation in liver of chick embryo". He was awarded Ph.D in the year 1976. He joined Bolpur College as a lecturer in Zoology and then in Bankura Sammilani College. In the year 1978 he joined to serve the Central Silk Board at Central Sericultural Research and Training Institute, Berhampore (W.B) and became section-in-charge of Silkworm Physiology. He was deputed to Peoples Republic of China in 1994 at Zhejiang Agricultural University,Department of Sericulture, Hangzhou for higher studies in Silkworm Physiology. In 1998 he joined the ZSSPO, Malda (WB) under National Silkworm Seed Organisation (CSB). In 1999 he joined SSPC, Mysore as an In-charge Officer. At present he is posted as Scientist D in Central Sericultural Research & Training Institute, Mysore. He had participated in a number of National and International meetings, seminars and symposia etc. He has acted as a referee for reviewing of a number of research projects on Sericulture. He is associated with many learned societies across the country. He has 53 research papers to his credit and as a lead author published in various Scientific Journals in India and abroad.

Silkworm Physiology

A Concise Textbook

Dr. Mohammaed Shamsuddin
Central Sericultural Research & Training Institute
Srirampura, Manandavadi Road
Mysore - 570 008, Karnataka

2018
Daya Publishing House®
A Division of
Astral International Pvt. Ltd.
New Delhi - 110 002

First Published, 2009
Reprinted, 2018

ISBN: 978-81-7035-623-3 (Hardbound)

Publisher's note:

Every possible effort has been made to ensure that the information contained in this book is accurate at the time of going to press, and the publisher and author cannot accept responsibility for any errors or omissions, however caused. No responsibility for loss or damage occasioned to any person acting, or refraining from action, as a result of the material in this publication can be accepted by the editor, the publisher or the author. The Publisher is not associated with any product or vendor mentioned in the book. The contents of this work are intended to further general scientific research, understanding and discussion only. Readers should consult with a specialist where appropriate.

Every effort has been made to trace the owners of copyright material used in this book, if any. The author and the publisher will be grateful for any omission brought to their notice for acknowledgement in the future editions of the book.

Published by : **Daya Publishing House®**
A Division of
Astral International Pvt. Ltd.
– ISO 9001:2008 Certified Company –
4736/23, Ansari Road, Darya Ganj
New Delhi-110 002
Ph. 011-43549197, 23278134
E-mail: info@astralint.com
Website: www.astralint.com

Printed at : **Replika Press Pvt. Ltd**

Dedicate to

My Parents

Late Mr. Mozammel Haque

and

Late Mrs. Jinnat Bibi

and

To my beloved

Teachers and Students

Acknowledgements

I am grateful to a large number of very competent names who have helped me in the preparation of this book. I am indebted to them for their keen interest and friendly co-operation which have rendered this book presentable. I am especially thankful to my wife Mrs.Karabi Chowdhury my son Ripon Md. Chowdhury and my younger borther Dr. Md. Mohiuddin for their genuine interest and constant encouragement for jotting down this book.

I am also grateful to my teachers especially to my beloved teacher Late Prof. Jitendranath Medda for his guidance, constant support and inspiring me during my student and research life. I am also thankful to Prof. Subrato Roy, former Dean of the Faculty of Science, Department of Zoology, University of Burdwan, Golapbag, Burdwan (WB) and President of the Zoological Association of Burdwan for his critical reading of the entire draft, suggestions and pain taking corrections. I also acknowledge with deep respect the apt and excellent foreword that he has written. I am also highly obliged to Dr. Taraknath Seth, former Head of the Department of Zoology, Burdwan Raj College, Burdwan, (WB). I am deeply indebted for the skill and support of my publisher Mr. Anil Mittal, without whose help this book would not have been possible to present it in its current form.

Dr. M. Shamsuddin

Foreword

Sericulture, which has a colourful history has recently been introduced as an important curriculum of study in both undergraduate and post-graduate levels of several Indian Universities. Due to tremendous applied and academic importance this discipline requires a coherently integrated and holistic approach to make the study a complete and effective one. But unfortunately there is dearth of hooks dealing solely with some aspects of this curriculum. Silk worm physiology has now been recognized by scientists and academicians as an integral and important part of this discipline that requires elaborate and extensive studies. But uptil now no serious attempt has been made by Indian authors to tackle this subject in a comprehensive way. This book may be considered to be a pioneer work in the field of silkworm physiology and will be able to arouse a sense of aptness of the students for the subject and will also be able to enhance the curiosity for its material contents.

The author Dr. Md. Shamsuddin, a product of Burdwan University, West Bengal is equipped with sound knowledge on diverse disciplines of sericulture. He was associated with Central Sericultural Research and Training Institute as a Scientist for more than thirty years and served different centers of CSR&TI with utmost seriousness and sincerity. The vast experience of the author has prompted him to write this book in which all ingredients of silkworm physiology have been presented in a scientific and methodical way. As an entomologist, I personally feel that this book will cater the need not only of the students of sericulture but also of other discipline of biological sciences. It will also be of some use to the teachers and researchers in the field of sericulture.

Subrata Roy 5/12/05.

Dr. Subrata Roy
Retired Professor of Zoology,
Formerly Dean, Faculty of Science
University of Burdwan,
Burdwan – 713 104, West Bengal

Preface

It is a reality that students at the undergraduate and post-graduate level of sericulture do not have a text book dealing in a concise and simple way with the physiological processes of silkworm. The main aim of this book is to cater to the students of sericulture up-to-date information on the subject of physiological aspect of silkworm.

In compiling this book I have tried my level best to reflect the contemporary outlook on the subjects limiting myself within the framework of the prescribed syllabi of most of the Indian Universities. All chapters of the book form a comprehensive introduction on silkworm physiology with emphasis on its morphology and anatomy as well.

The special feature of this book includes comprehensive text, written in a simple and easily understandable language, elaborate and self-explanatory figures and diagrams.

I am aware of my shortcomings and difficulties faced by me as an author to publish such a complex book single handedly. Any error or oversight is strictly my own. I would request the readers not to hesitate in pointing out the mistakes so that they can be removed and edited in the next edition. I would further request the readers to send their suggestions or remarks regarding this book to the publisher, which would be highly appreciated.

Dr. M. Shamsuddin
E-mail: shams.mohammed@yahoo.co.in

Contents

Chapter 1

Systematic Position of Mulberry and Non Mulberry Silkworm and their Salient Features

Kingdom: Animal Kingdom or Animalia

1) Absence of cellulose layer on cell layer
2) Responds towards irritability
   (Sense organs well developed)

Sub Kingdom: Non Chordate or Non Chordates

1) Absence of notochord or vertebral column
2) Ventral nerve cord present
3) Absence of gill slits

Phylum: Arthropoda

1) Presence of joint appendages

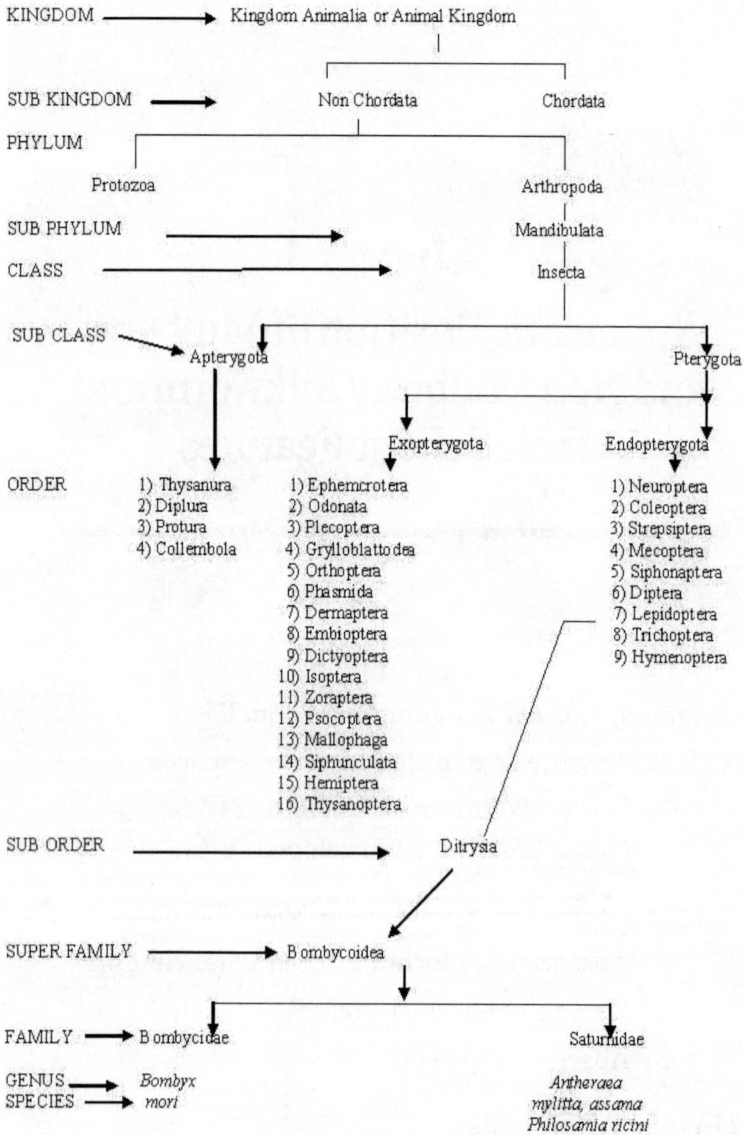
KINGDOM
Kingdom Animalia or Animal Kingdom
SUB KINGDOM
Non Chordata
Chordata
PHYLUM
Protozoa
Arthropoda
SUB PHYLUM
Mandibulata
CLASS
Insecta
SUB CLASS
Apterygota
Pterygota
Exopterygota
Endopterygota
ORDER
1) Thysanura
2) Diplura
3) Protura
4) Collembola
1) Ephemcrotera
2) Odonata
3) Plecoptera
4) Grylloblattodea
5) Orthoptera
6) Phasmida
7) Dermaptera
8) Embioptera
9) Dictyoptera
10) Isoptera
11) Zoraptera
12) Psocoptera
13) Mallophaga
14) Siphunculata
15) Hemiptera
16) Thysanoptera
1) Neuroptera
2) Coleoptera
3) Strepsiptera
4) Mecoptera
5) Siphonaptera
6) Diptera
7) Lepidoptera
8) Trichoptera
9) Hymenoptera
SUB ORDER
Ditrysia
SUPER FAMILY
Bombycoidea
FAMILY
Bombycidae
Saturnidae
GENUS
SPECIES
Bombyx
mori
Antheraea
mylitta, assama
Philosamia ricini

2) Body is metamerically segmented and covered with non-living thick chitinised cuticle forming the exoskeleton
3) Moulting or ecdysis is found during the life cycle
4) Presence of compound eyes with mosaic vision
5) Perfect bilateral symmetry
6) Circulatory system open and heart placed on the dorsal side of the alimentary canal
7) Development may be direct or indirect
8) Cilia or flagella are completely absent
9) Excretion occurs by malphigian tubules

Sub Phylum: Mandibulata

1) Terrestrial, fresh water or marine
2) Respiration by gills, tracheae or skin
3) Antenna 1 or 2 pairs
4) Mandibles one pair
5) Walking legs 3 or more pairs

Class: Insecta

1) Terrestrial arthropods with tracheal mode of respiration
2) Body distinctly divisible into head, thorax and abdomen
3) Antenna 1 pair present
4) A pair of mandibles and two pair of maxillae present.
5) 3 pairs of thoracic walking legs
6) 1 or 2 pairs of wings present

7) Excretion by malphigian tubules
8) Sexes are usually separated
9) Genital opening is situated near the posterior end of the body
10) Post embryonic development is rarely direct and a metamorphosis usually occurs

Sub Class: Pterygota

1) Adults are winged, sometimes secondarily apterous
2) Metamorphosis is complete or incomplete.
3) Adults without pre genital abdominal appendages.
4) Adult mandibles, unless highly modified, articulating with head-capsule at two points.

Division: Endopterygota

1) Wings develops internally
2) Metamorphosis complete in the resting or pupal stage.
3) Larvae are highly specialized, which differ from adult in structure and habit.

Order: Lepidoptera

1) 2 pairs of membranous wings with few cross vein.
2) Body wings and appendages covered with broad scales
3) Mandibles almost always absent or vestigial and the principle mouth parts are generally represented by suctorial proboscis formed by the maxillae
4) Larva eruciform
5) Pupae usually more or less object and generally enclosed in a cocoon or an earthen cell

6) Wing tracheation complete. Forewings always larger than hind wings
7) Complete metamorphosis.
8) Typical caterpillar or larva with biting and chewing mouth parts. 3pairs of thoracic legs, 5 pairs of abdominal legs designed as pseudopods and often with silk gland.

Sub Order: Ditrysia

1) With copulatory pore on sternite VIII and eggs pore on sternite IX
2) Fore wing without jugum or fibula and the hind wing with frenulum.
3) Male sternite IX usually U- or V-shaped, often produced into a large saccus.
4) Larva with not more than 5 pairs of crochet-bearing abdominal legs.
5) Pupa obtect, without functional mandibles.

Super Family: Bombicoidea

1) Maxillary pulps and tymponal organ absent
2) Frenulum atrophied
3) Proboscis rarely developed
4) Antennae pectinate specially in males
5) Cu_2 absent in both wings, hind wing with Sc + R_1 usually diverging from cell and R_s, or only connected with cell by a cross vein (R_1).
6) Chaetosema absent.

Family: Bombycidae

1) Proboscis absent
2) Pectinate antennae in both sexes.
3) Larvae are of 2 types male and female
4) Hind wing with Sc+R1 connected with cell by a cross vein, fore wing with M1 free or shortly stalked on Rs

Genus: *Bombyx*

1) Larva are glaborous and elongate, usually with a mid dorsal horn on the 8th abdominal segment
2) They form dense silkworm cocoon

Species: mori

Commonly known as silkworm with 100 per cent domestication. It differs from its allied species in its specific brooding behaviour. Moultinism which mainly depends on the climatic as well as genetic endowment of the organism. The only natural food of this species is mulberry leaf.

Family: Satrernidae

1) Antennae are prominently bipectinate in both sexes. The ramy being largest in male.
2) The labial pulps are minute and there is no frenulum
3) Presence of transparent eye spot near the centre of each wing
4) Hind wing with vein Sc+r1 diverging from cell base. M2 arising at or in from of middle cell nearer M1 than Cu, Cu2 and frenulum completely absent.

Genus: ***Antheraea***

1) The worms are polyphagous

Species: mylitta

1) A long prominent peduncle with ring is present at the cocoon
2) Centrally placed prominent eyes and transparent spot on the wing.

Species: assama

1) Larvae are green in colour with thorny warts, more prominent in early stages.
2) Cocoons are usually formed in holes or prefer the space between two lids.
3) Muga eggs are brownish green in colour and chitin is cellular in structures which differ from other ***Antheraea*** species.
4) Brown cocoons with very rudiment peduncle.

Genus: ***Phhilosamia***

1) Moth is somewhat brownish green in colour and possesses narrow crescents on the wings instead of usual central eye spot.
2) Body of the worms contain 6 pointed fleshy thorns
3) Peduncle is absent in the cocoon
4) Cocoons are open mouth in nature.

Species: ricini

1) Amount of scales in the abdomen is less.
2) Cocoons are loose and white.

Species: cynthia

1) Amount of scales in the abdomen is more
2) Cocoons are compact and light brown in colour.

Chapter 2

External Morphology, Silkworm Eggs, Larva, Pupa and Adult

1. Shape and Weight of Eggs

In general the silkworm egg is a small, flat, ellipsoid, but the shape and size are variable according to races and individuals. At the one end of the ellipsoid there is a micropyle, the entrance of the spermatozoa. By upper view, either the left side or the right side of the micropyle it is somewhat upheaved, accordingly, the shape of the egg is not completely symmetrical.

An egg weighs about 0.6mg and shows approximately 1.075 in specific gravity. Generally the egg of European races is large and heavy, while that of tropical polyvoltine races is small and light; and among the eggs laid by a mother moth, the eggs, which are laid earlier, are heavier than those which are laid later.

2. Colour of Eggs

Newly laid eggs are light yellow in colour, and the colour of the non-diapause ones does not change up to the head pigmentation stage. Usually the hibernating eggs show their peculiar colour in three or four days after deposition, being non-diapause eggs, as if they were diapause ones. Accordingly, it is said that the pigmentation of eggs coincides not always with the diapause of them.

As to the colour of the hibernating eggs, with the exception of white eggs, which are brown or red eggs, which are rather exceptional cases, there is a variation in the egg colour of common silkworm races, such as deep purplish gray, dark purplish gray, pinkish gray, brownish gray, light brownish gray etc. Generally deep coloured eggs are found in Japanese races, while eggs of Chinese races are commonly light in colour.Greenish appearance of the eggs (egg shell). The light yellow colour of chorion turns darker when it is affected with ammonia gas in the same way as the pigment of light green cocoon does. The colour of eggs also changes by season.

For instance, the colour is deep in summer, because the pigment scatters evenly throughout the serosal cell. On the contrary in winter the pigment is located at one corner of the cell, making the colour of eggs lighter, until next spring when it scatters again and the colour of the eggs turn deeper.

3. The Gluey (Adhesive) Substance on the Surface of Eggs

Ordinary silkworm eggs receive gluey substance only on the underside of them during deposition and adhere to the egg card. However, some eggs, such as Bagdad races, have no gluey substance; therefore, those eggs cannot adhere to the card.

At the time of laying eggs the gluey substance is rhythmically discharged from the mucous gland opening at the upper portion of the oviduct of the female moths and flows down on the ink-pad like tissue full of wrinkles, locating at the lower side of the oviduct. This gluey substance is only given to the underside of eggs by the way of laying eggs from the ink-pad like tissue of oviduct.

The gluey substance can be found only on the one side of the loose formed eggs when they are stained with red ink or things like that, because the loose formed eggs were washed down in water after they are deposited on starched paper. When pupae or female moths are kept at high temperature or stored in a cold room the gluey substance is put not only on the under-surface of eggs, but, on the upper surface of them, making adhesion of eggs irregular. The old practice that the regularly deposited eggs are preferred is reasonable, because the eggs laid in unfavourable conditions can be confirmed by the form of adhesion of eggs.

4. Chorion (Egg Shell)

The eggs are covered with chorion. The chorion occupies about 10 per cent of the eggs in weigh and consists of chorionin, which is different from chitin, the major component of insect skin, rather resembling Karatin (protein of wool), and contains much sulfer and nitrogen. The egg has a micropyle at an end, the micropyle branches off usually three canals soon inside the entrance, passing through the chorion. Surrounding the micropyle, there is a chrysanthemum-like marking on the chorion, being somewhat upheaved and projecting outward. The whole surface of chorion, excepting the chrysanthemum marking, is covered with a net like pattern, and numerous needles–shaped respiratory canals are distributed over the same area of the chorion. The net like pattern is the print of the shape of follicle cells which secrete the chorion in the ovary of mothers showing characteristic types according to races.

The chorion is about 0.02mm in thickness showing racial difference. That is it is thickest in European races followed by Japanese races and Chinese Univoltine races, being thinnest in Chinese bivoltine ones. Even in the same egg the thickness is different locally. In the upper view of the external form of an egg, the curvature of the right side is different from that of left at the end where the micropyle exists, that is, it is asymmetrical, and the chorion layer with smaller curvature is thinnest, from there the newly hatched larva comes out eating that part of chorion.

At the time of conducting the hydro-chlorization of eggs, if ferric chloride or ferrous chloride is added to acid in the rate of 1-2 per cent in weight, the chorion of some eggs treated with acid may be stained in green or yellow colour. The chorion of eggs of Chinese races is stained in most cases, but that of Japanese and European races mostly cannot be stained by this method.

Internal Morphology of Silkworm Eggs

Beneath the chorion, there is a very thin vitelline membrane and in the next place the serosa which consists

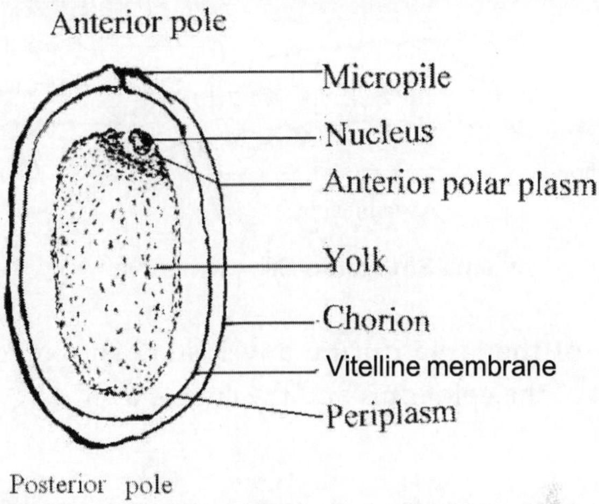

Structure of Silkworm Egg

of cells arrange in one layer as paving stones and is rich in pigment granules. Being wrapped by those two membranes, semiliquid yolk fills the egg and an embryo, which is to grow into a larva, can be seen on one side of yolk mass.

Larva

The silkworm larva when newly hatched is black or dark brown. It has a large head and the body is densely covered with bristles so that it looks like a hairy caterpillar. There are four pairs of tubercles; the sub dorsal, supra spiracular, infra spiracular and basal tubercle each of which carries three to six setae. As the larva grows, it becomes smoother and lighter in colour due to the rapid stretching of the cuticular skin during, the different instars of the larval stage. The larval body is composed of the same segments as any insect, head, thorax and abdomen. The entire body is covered with a thin and elastic chitinous cuticle which is capable of being extended considerably to permit rapid

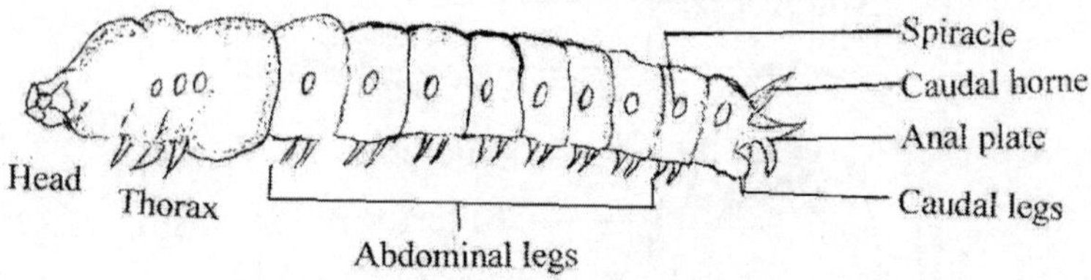

External Structure of Silkworm Larva

growth of the larva during any inster. The integument consists of the epidermis and the cuticle over it.

Head

The head consists of six body segments fused together with a cranium. The second, fourth,fith and sixth segments carry appendages which are modified into antennae, mandibles, maxillae and labium respectively. Median epicranial suture is well developed and prominent. Similarly

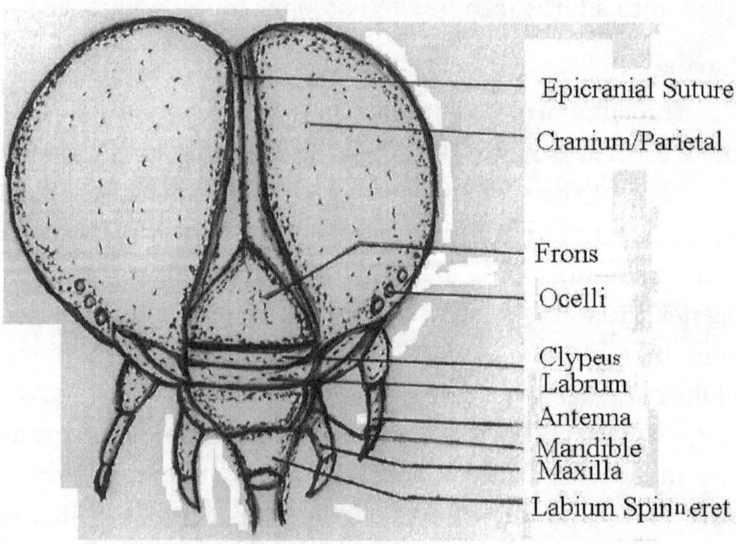

Head

on the outer side the clypeus and the labrum are also prominent. There are six pairs of ocelli or larval eyes which are located behind and a little above the base of the antennae.There is a pair of antennae formed of five jointed segments and these are used as sensory organs (feelers). The mandibles are well developed for mastication.

Antenna

The antenna is formed of five segments.Antacoria is joined to the head capsules by antennaria.

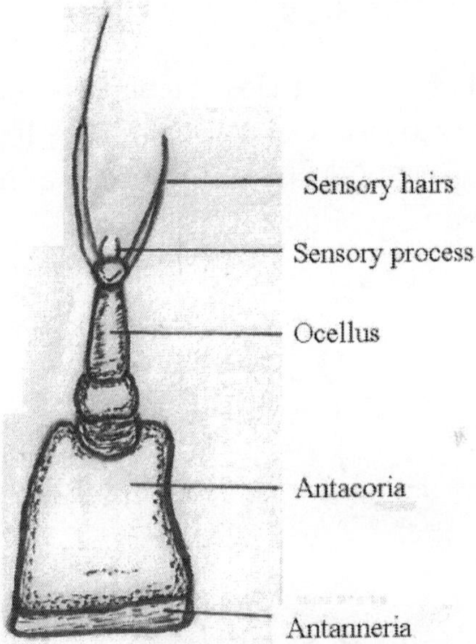

Antenna

Mandible

The Masticatory appendage of the oral somite, a greater part being formed of a blakish hard chitin layer. On the upper side of the outer edge there is the accetabulum and on the lower side the condyle.

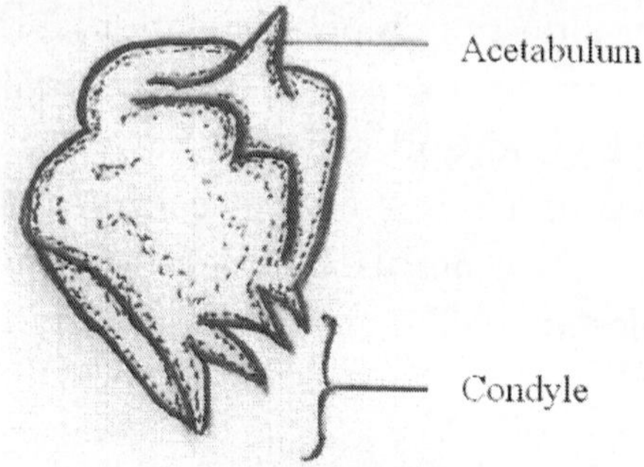

Mandible

Maxilla

The maxilla is situated close behind the mouth on the ventral side. It is consist of cardo,stipes, maxillary lobe and maxillary palpi. Maxillary lobe and maxillary palpi discriminate the taste of food.

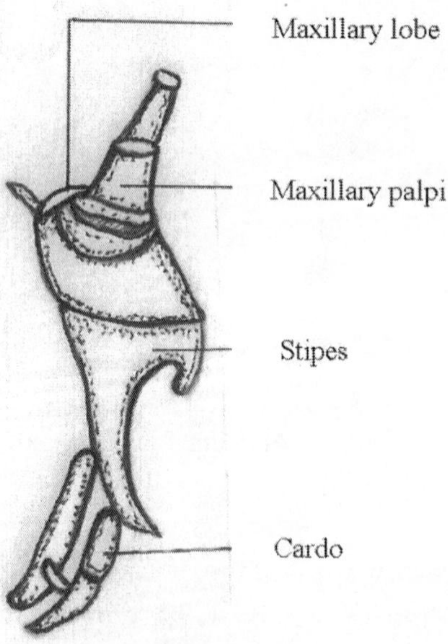

Maxilla

Labium

The labium (lower lip) is located ventrally carrying a big sized, lightly chitinized mentum,the posterior dorsal edge is white and soft sumentum and the successive chitinous blak part is prementum. Distally the prementum carries a median process or spinneret through which silk is expelled from the silk gland to form the silk bave or thread. The sensory labial palpi are found on both sides of the spinneret.

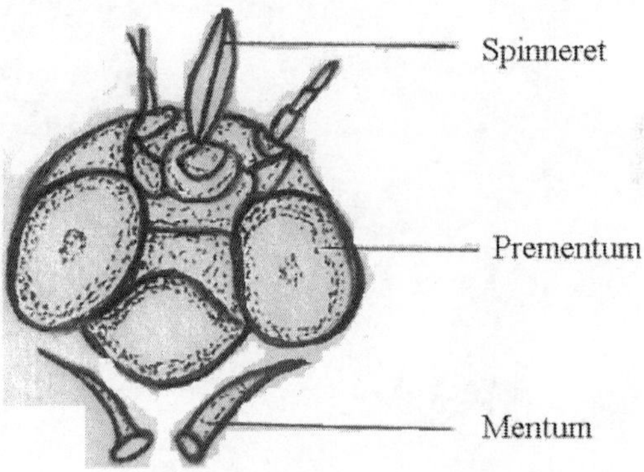

Labium

Labrum (Upper Jaw)

The dorsal side joins to the frons, clypeus and covers a part of the mandible. On the outer side six pairs of setae and on the inner side three pairs of processes present.

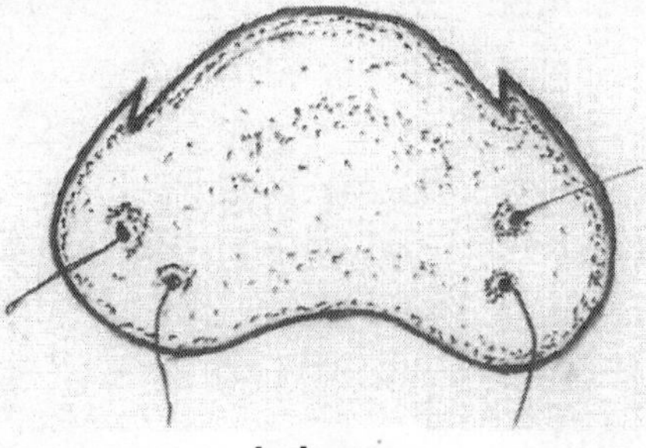

Labrum

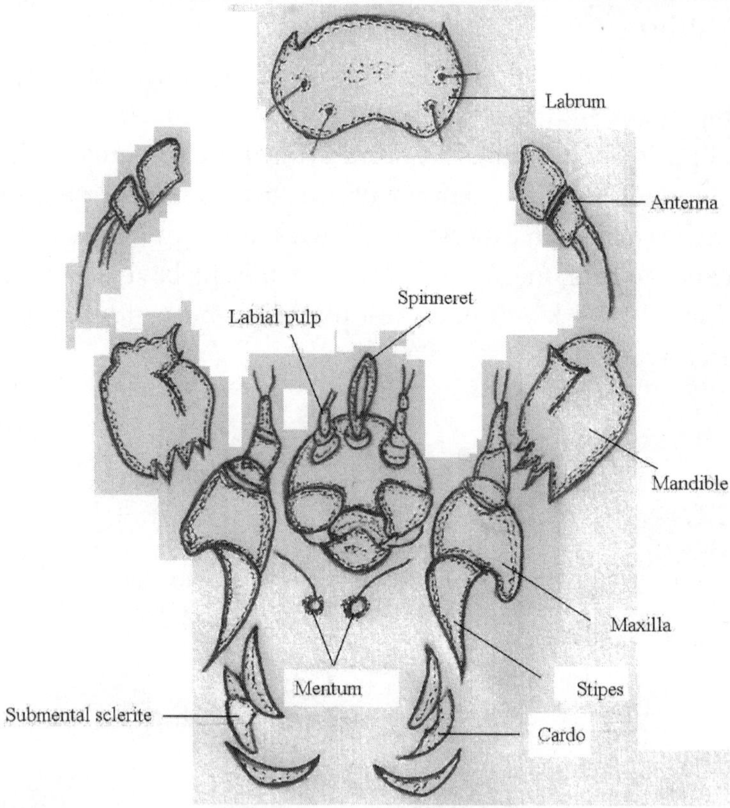

Mouth Parts of Silkworm

Thorax

The thorax consists of three body segments called the pro-meso- and meta- thorax.Each of the three thoracic segments carries ventrally a pair of legs each comprising in turn three jointed segments. These are the true legs which are conical in shape and carry sharp distal claws. These claws are not used for crawling but for holding mulberry leaves while feeding. There is an eye-spot in the dorsal side of the second thoracic segment on all silkworms except plain pattern silkworm.

Abdomen

The abdomen is comprised of eleven body segments although only nine can be distinguished as the last three are fused together to form the apparent nineth segment, the anal plate and the cadual legs. The third to sixth and the last abdominal segments each bear a pair of abdominal legs which are fleshy, unjointed muscular protuberances. At the extremity they form a sort of disc with a series of hooks inwardly curved and arranged in a semicircle. On the dorsal side of the eighth abdominal segment, the larva carries the caudal horn.

The abdominal segments carry the sexual markings which develop distinctly in the fourth and fifth instars in the eighth and ninth segments on the ventral side.

In the female the sexual markings appear as a pair of milky white spots in each of the eighth and ninth segments.The pair of spots on the eighth segment is known as Ishiwata's fore glands and the pair of the ninth segment

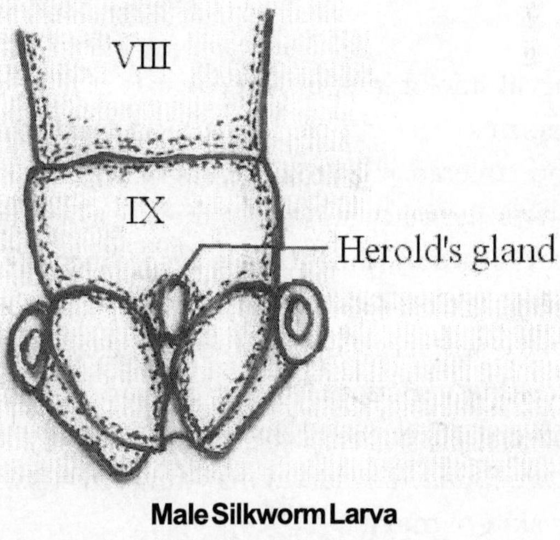

Male Silkworm Larva

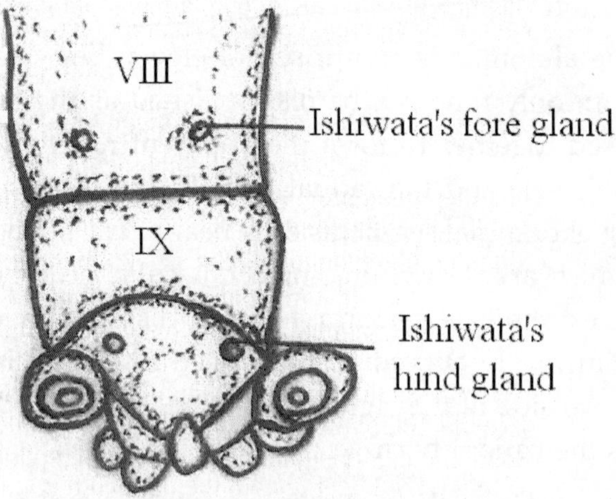

Female Silkworm Larva

is referred as Ishiwata's hind glands. In the male a small milky white body (Herold's gland) appears at the centre of the ventral side between the eighth and ninth segments.

There are in all nine pairs of spiracles placed laterally on either side of the body. They are found on the first eighth abdominal segments. These are the breathing or respiratory pores.

The larval skin or integument consists of the cuticle and the hypodermis. The cuticle is made of chitin as well as protein and covered with thin layers of wax. Nodules are found all over the body surface of the silkworm larva. The distribution of these nodules differs according to the variety of silkworms, being less in number in varieties having lustrous body surfaces, such as the Chinese bivoltine and the multivoltine varieties. The nodules are found in large numbers overparts of the body where normally the larval markings are found. The larval markimgs in silkworms are caused by skin pigment.

Pupa

The pupal stage is generally called the resting and inactive stage in silkworm when it is incapable of feeding and appears quiescent. This is a misnomer. The pupal stage is a transitional phase during which definite changes occur. During this period the biological activity of the larval body and its internal organs undergo a complete change and assume the new form of the adult moth. The mature silkworm larva passes through a short transitory stage of pre pupa before becoming a pupa. During the pre-pupal stage the dessolution of the larval organs takes place and this is followed by the formation of adult organs during the pupal stage. Soon after pupation the pupa is white in colour

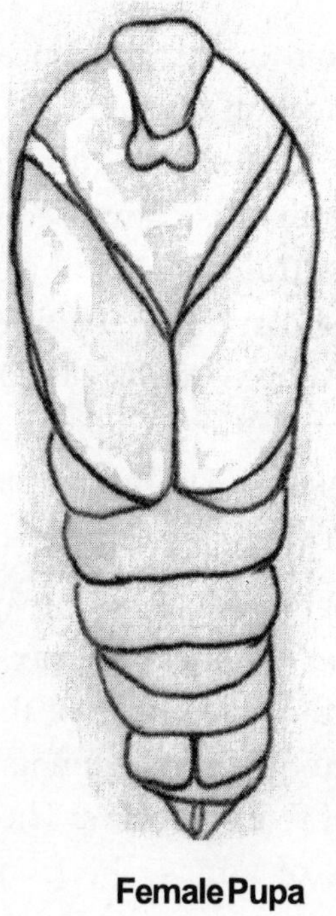

Female Pupa

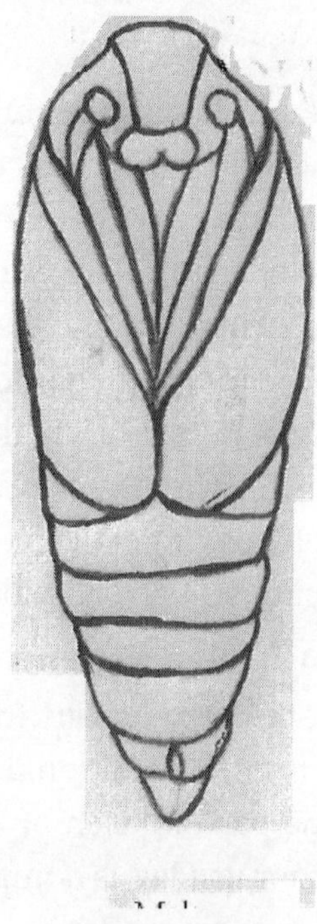

Male Pupa

and soft but gradually turns brown to dark brown and the pupal skin becomes harder.

The prominent morphological parts visible are a pair of large compound eyes, a pair of large antennae, fore and hind wings and the legs become large. Ten of the abdominal segments can be seen on the ventral side, but only nine of them are visible on the dorsal side. Seven pairs if spiracles are found in the abdomen. The last pair is nonfunctional. Sex markings are prominent and it is much easier to determine sex in the pupal stage than of the larval stage. The female has a fine longitudinal line on the eigth abdominal segment, where as such a marking is absent in the case of male. The male has only a round spot.

Adult

The adult moth emerging from the pupa is incapable of flight as a result of its domestication for more than five thousand years. It is not feed during its short adult life. The body of the moth like that of the larva is composed of the three distinct segments: head, thorax and abdomen. Another important characteristristic in common with other moths and butterflies is that of the body surface is covered with scales. The compound eyes are situated on either side of the head. The ocelli are absent. The antennae are conspicuous, large and bipectinate.

The thorax consists of three segments namely pro- meso- and meta thorax as in the larva. The meso thorax is the largest and is pentagonal. There are three pairs of thoracic legs, one pair on each of the three thoracic segments. Each of the thoracic leg is composed of five segments. The meso and metathorax bears two pairs of wings, the front pair over lapping the hind pair when the moth is in the resting position.

In the male eight segments are visible; in the female seven. There are six pairs of spiracles present laterally on either side of the body.

Morphologically the female and male can be easily distinguished in the adult stage. The female has comparatively smaller antennae, its body and the abdomen are fatter and larger and it is generally less active than the male moth. At the caudal end the male moth has a pair of hooks known as harpes where as the female moth has a knob-like projection with sensory hairs. These differences help to a large extent in separating the sexes for preparation of hybrid eggs.

Chapter 3
Integument

The integument, or body wall of silkworm is the exterior tissue of the body and appendages, it originally comes from ectoderm. The integument comprises of the basement membrane, epidermis and cuticle. The cuticle is the exterior layer. It is produced by epidermal cells, and covers the entire body. Body wall provides physical protection for internal organs and reduces water loss to a very low level, prevents entry of foreign materials such as pathogen and other detrimental things, both living and non living, into the body. It acts as receptors of foreign stimuli, and plays important role in energy exchange and production of body colours. It serves as a framework of body forms and acts as a metabolic reserve to be used cyclically to construct new integument in the next stage, or during periods of metabolic activity or starvation. For instance, interior portion of the cuticle, the endocuticle will be digested and reabsorbed during ecdysis.

Its apodemes, which are formed by the invagination of the integument, provide internal area for muscle attachment.

Structure and Function of Integument

Larval Integument

The larval integument comprises of three main layers, the exterior layer is the cuticle, the middle layer is the epidermis and the interior layer is the basement membrane.

Basement membrane is an acellular multilayer membrane under electron microscope. It is thought that basement membrane may be produced by haemocytes, though epidermal cells also participate in its synthesis during healing of the wound. Its main component is neutral mucopolysaccharide. This membrane function as an isolation of epidermal cells from body cavity, it permits selectively chemical compounds and hormones enter epidermal cellular layer.

Epidermal layer is rested on the outer side of the basement membrane, and is one cell thick layer. During growing period, epidermal cells are cuboidal, its nucleus is slightly spherical, its endoplasmic reticulum and golgi body are well developed and plasmic filaments arise from the apices of the cell, these filaments form the passage extending into the cuticle. During the formation of new cuticle, the cellular membrane forms the longitudinal folds in order to increase the absorbing area. Sometimes the epidermal cells contain pigments and urate crystal. The epidermis is only living layer, it secretes new cuticle under stimulation of moulting hormone at various developmental phase, causes periodical ecdysis. And the epidermal cells may be differentiated into sense organs and various integumental outgrowths such as setae and scales etc,and may be modified into various glands.

The cuticle is produced by epidermal cells; its main components include water, protein, chitin and lipid. The cuticle includes three primary layers: the inner endocuticle, the middle exocuticle, and the outer epicuticle. The endo cuticle is rested on the epidermal layer, it is the thickest layer, and composed of lmellae, four fifths of the cuticle in thickness belongs to endocuticle, it is mainly composed of chitin and protein. Chitin and protein bind together to form a complex glycoprotein and chitino protein. Chitin is a kind of N-polysaccharide, its physical and chemical properties is similar to cellulose and it does not dissolve in hot water, diluted acid, diluted alkali, acetone, alcohol, trichloromethane and ether, but it can be extendable. Endo cuticular protein is a sort of dissolvable anthropodeal proteins which have their special property. Endocuticle is not tanned with quinines (converted phenols), thus it is softy and full of elasticity and flexibility. Endocuticle does not contain pigments and most of endocuticle will be dissolved and reabsorbed during ecdysis.

Exocuticle is very thin, its thickness is about one fifth of the cuticle, it is differentiated from endocuticle and composed mainly of chitin and protein. In exocuticle the protein is tanned into chitinous exoskeleton, in the newly exuviated larvae, newly pupae and newly adults, the body walls are softy and light colour because their integuments have not developed completely and tanned. But sometimes later, epidermal cells generate a kind of polyphenol oxidase which is transferred to epicuticle through the pore canals and accumulates right beneath the epicuticle, and reacts with tyrosine in the cuticle. The reactions forms 3,4 dihydroxy phenylalanine, called simply "dopa", which is transformed into N-acetyldopamine through dopamine. The production of N-acetyldopamine is primary component for protein tanning. The N-acetyldopamine is oxidated to form quinone, which diffuses and links protein molecules together. This process makes the soft and dissolvable protein tanned into hard scleritized protein. The tanning of the protein is usually accompanied by darkening (melanization) that is because the quinine produced may further polymerise into melanin. At this moment the exocuticle is formed and meanwhile the integument gets dark.

The epicuticle is the outermost layer of the cuticle and is an exceptionally thin region, which is composed of two layers: the lipoprotein layer and the wax layer; Lipid and protein is the main chemical element of the former (Lipoprotein layer). The lipoprotein layer is deep in colour, and no elasticity and no flexibility, and is firstly formed layer during production of new cuticle. It can isolate new cuticle from the old cuticle and possess selective permeability. It permits the products digested by moulting fluid to penetrate this layer, but it cannot be dissolved by the moulting fluid and does not permit the moulting fluid to penetrate this layer. In addition, it seems to be highly resistant to moulting fluid, and to protect the newly formed cuticle. The lipoprotein layer is covered with wax.

The wax layer exhibits well is lipophilicity and strong hydrophobicity, it can prevent foreign materials from entering the body and water in the body from overlosing.

The permeability of the integument to water depends mainly on the quality and quantity of the wax in the epicuticle. For instance, multivoltine varieties have fewer wax than the bivoltine varieties, therefore, the integuments for the multivoltine varieties are more permeable to water than those of bivoltine varieties, larvae reared with high quality mulberry leaves have thicker wax layer than those with young,seedling or old mulberry leaves. The wax amounts are not significantly different among various development stages except newly moulted silkworm, but after beginning to cocoon formation the amounts of wax are getting more and more, especially during pupation, the wax amounts increase sharply. The permability to water in late pupal stage is larger than that in early pupal stage. This is due to not only increasing the wax amount but also rising the wax melting point. There are tiny pore canals traversing the cuticle, their bases link with the epidermal

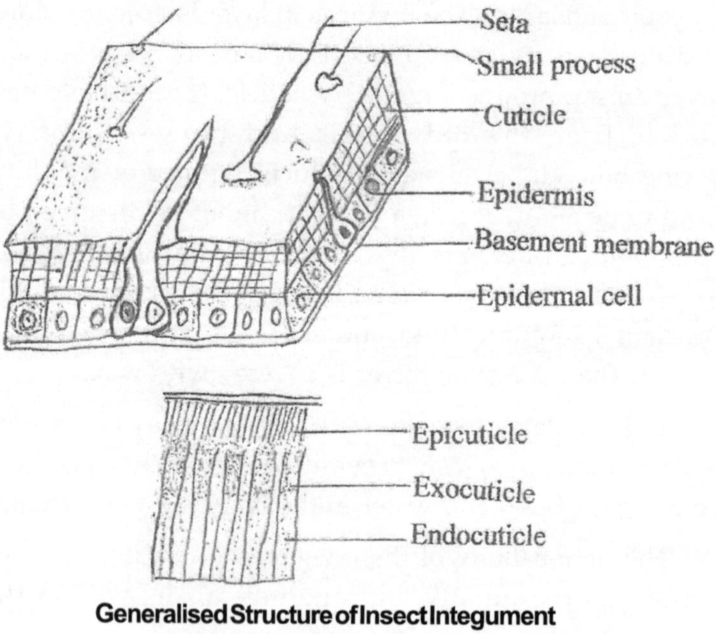

Generalised Structure of Insect Integument

cells, and their apices reach to inside small processes and branch. The pore canals function as transporting secretions such as chitin, protein and lipid by epidermal cells.

Being exoskeleton, cuticle bears muscles, these cuticle muscles are bound together in a special mode so that the force produced by the muscular construction can reach to the exoskeleton, but these two parts will depart and reconstruct new linkage, linkage between muscles and integument depends on the skin at the cuticular layer. These two components form tonofibrillae.

Pupal and Adult integument

The pupal integument is almost the same as larval integument, and is composed of basement membrane, epidermis and cuticle,too. But unlike larval cuticle the pupal cuticle is composed of four layers, that is epicuticle, exocuticle, mesocuticle and endocuticle. The mesocuticle layer is between exocuticle and endocuticle, and it is colourless, but it can be dyed by a special dye, purified from dyewood. In addition,the exocuticle is greatly tanned,and harder than that of larvae.

The adult integument is similar to pupal integument except covering by scales. The cuticle is composed of four layers too. That is, epicuticle, exocuticle,mesocuticle and endocuticle. However, new cuticle differentiation is much earlier for the adult integument than that for pupal integument. Usually, the day before adult emergence the tanned and coloured exocuticle can be seen, but tanning extent is much lighter than that in pupae. However, cuticular structure in adult and pupal stage may be different depending on various body regions. For example, the exocuticle and mesocuticle are absent at conjunctiva, thus in these regions, the cuticle appears to be softy and flexible.

Integumental Derivatives

The integumental derivatives are formed through modifications of epidermal cells or the cuticle. Setae, scales are common out growths of the integument; moulting gland is common derivatives inside the body wall setae.

Setae

There are many setae on the integument, setae are formed by an epidermal cell through outgrowth, this epidermal cell is called trichogen cell. The cytoplasm above nucleus firstly produce many granules, the trichogen cell accumulate these granules and meanwhile outgrow into a setae, outside the trichogen cell at the seta base is the thin tormogen cell, which forms a sheath for the trichogen cell. Many trichogen cells are linked with sensory cells beneath the epidermis, thus to form sensillae. During ecdysis, seta falls off at the base of seta; new setae are produced by the epidermal cells.

There are also papillary, spiniform and rod shaped apophysis on the larval integument, but these apophysis differ from the setae in structure.

Scale

Entire body of the adult is covered by the scales. Scales and setae are homogeneous, their structures are capsular when forming, and the out growths are contracted finally into scales. The base of scale appears to be a handle. The top margin bears many saw like teeth. But some scales look like hairs, these scales have no saw like top margin, or few saw like teeth on the margin. Generally speaking, the scales on the wings are larger and shorter, with more saw like teeth on their top margin. The scales on the body are thinner and longer, with less saw-like teeth on their top margins. There are more scales at the anterior portion of each segment than at the posterior portion. There are more scales on the dorsal and on the lateral surfaces but there are no scales on the conjuctiva.

Moulting Gland

The moulting gland is also called epidermal gland, it is formed by the modification of the epidermal cells. The moulting gland exits in the epidermis, and appears to be capsular. Each moulting gland is composed of three cells; the largest cell at the centre is the secretory cell, which has a branchy nucleus. Two lateral sides of the secretory cell are two small conducting tubular cells. Two cells combined together to form the tubule, and have round nucleus. There are fifteen pairs of moulting glands in the larval integument.Each of three thoracic segments, and the first to ninth abdominal segments has one pair of moulting glands, respectively on the lateral surfaces, other three pairs of moulting glands exist at the base of the abdominal legs. These moulting glands have special secretary functions. The secretary cells can secrete the moulting fluid, which contains chitinase and proteinase. The moulting fluid is excreted into the space between the newly forming epicuticle and the old endocuticle by the secretary cells. It dissolves the endocuticle of the old cuticle, and it functions as lubrication. The moulting glands are different in size, the moulting glands in the abdomen are larger than those in thorax. At the very beginning of moulting, because of the secretion of the moulting fluids, while during the ecdysis, the glands recover gradually with excretion of the moulting fluid, formation and thickening of the new cuticle, and then the moulting glands become inactive again.

Chapter 4
Muscular System

The various movements in silkworm occur depending upon the contractility of the muscles in the body. Muscles are arranged segmentally. However, their contractile action is controlled by nervous system.

Muscular Arrangement and Histological Structure

The muscle derives from the mesodermal cell. According to their position and range they affect, the muscles can be arranged in three categories: Skeletal muscles, visceral muscles and skeletal visceral muscles.

The skeletal muscles attach themselves to the integument, and are composed of parallel muscular fibres. They are widely distributed and control the movements of both partial and entire body. The muscles in the head control mainly the movements of the appendages. One ends of the muscles arise from the tentoria, the other ends attach to the appendages or mouth parts. These muscles include antennal muscles labral muscles mandible muscles maxillary muscles and labial muscles etc. The muscles in thorax and abdomen can be divided into dorsal muscles and ventral muscles by the longitudinal tracheal line. Both the dorsal and ventral muscles comprises outer,middle and inner layer. The muscles in the outer layer are short and thin, almost all are arranged obliquely, they control the contractile and flexing movements of the entire body as well as the movements of respective appendages. The muscles in the middle layer are slightly longer than those in outer layer. They are arranged longitudinally or obliquely, and can cause the body laterally flection. While the muscles in the inner layer are larger and long most of them are arranged longitudinally, they can cause the body to extend and shorten and cause the flexing and crawling movements of the body.

The visceral muscles are covering on the outer surface of the visceral organs. Some of them are arranged longitudinally or circularly, while the others are arranged abnormally to form network of muscular layer. These visceral muscles control the movements of the various visceral organs.

In the skeletal visceral muscles, their one ends arise from the integument, while the other ends attach to the visceral organs. Some of the muscles of the digestive tract and on the silk glands are of this type. These muscles play important role in fixing the inner organs.

Like in other insect the muscles in silkworm comprises many parallel muscular fibres which are all cross-striated although in some visceral muscles the striation may be

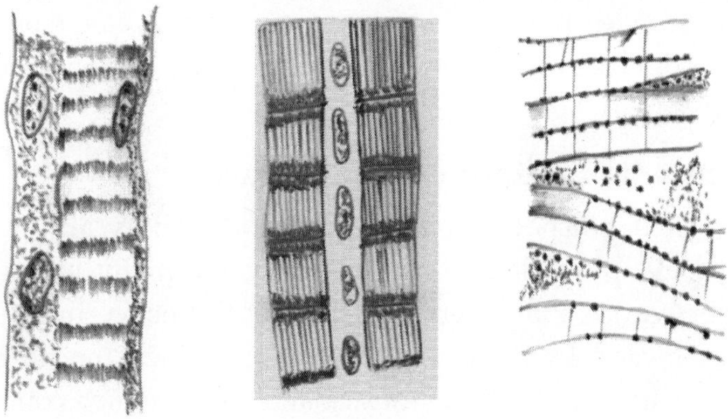

Sarcolemma Sarcoplasm Sarcosome

Different Types of Muscles

difficult to detect. Each muscular fibre is a larger multinuclear cell, which is composed of sacrosome. The sarcolemma is elastic and it covers on the outer surface of the muscular fibre. The myonuclei are cylindrical or oval, with obvious nucleoli. The sarcoplasm in the muscular fibre are comparable to the cytoplasm in other cells, are full of materials such as glycogen, fat and so on and the cell organelles such as Golgy bodies, endoplasmic reticula and granular mitochondria and the numerous myofibrilla grouped in clusters.

Musculature in Adult

The musculature in adult is similar to that in larval stage. Because the mouth parts has degenerated greatly in the adult, the corresponding muscles in the head are reduced, however, antennal muscles are well-developed. The muscles in the thorax are developed much better than those in the larvae. Almost all the thorax is filled up with the muscles.

The muscles in the abdomen are mainly those arisen on the reproductive organs, they play important role in mating and egg laying.

Muscular Excitation and Contraction

Excitability, conductivity and contractility are of the muscular physiological characteristics. Normally, muscular fibers can excite after they receive the nerve impulses transmitted from the motor neurons. When the normal association between the muscle and nerve, which controls the muscle is broken, the muscle will lose the ability to move. The muscles have the ability to conduct the excitation. No matter where the excitation occurs in a muscular fibre, the excitation will conduct along the entire fibre. However, the conduction of the excitation is limited in the fibre. This characteristic is the prerequisite for fine regulation of the nervous system to the muscular contraction. When the excitation reaches to some extent, the muscles start contraction.

Motion is produced by the action of muscular contraction in certain strength, speed and order. When some of the muscles contract, the other muscles may be in relaxation. When in the counter-movement, contracting muscles relax, while the relaxed muscles contract. It is these muscular alternative movements step by step that they result in the movements of some parts of the silkworm. The larval crawling, pupal abdominal rotating, adult wing vibrating are all resulted from the alternative muscular contraction.

Chapter 5
Silk Gland and Silk Biosynthesis

In Lepidopteron larvae the labial glands are transformed into organs for producing silk, utilized in the formation of larval shelters and cocoons. Silk gland is ectodermal in origin and lie on the lateral side of the gut. It is a paired structure cylindrical tubular organs of exceedingly variable length with characteristically branched nuclei.

Silk gland is divided into 3 distinct parts

Anterior Part, Middle Part and Hind Part or Posterior Part

Anterior Part

It is narrow slender, 2cm long. In this part 250 cells are present. Anteriorly two anterior parts unite and a pair of Fillipi's gland opens on it. At the end of the anterior portion a spinneret is present.

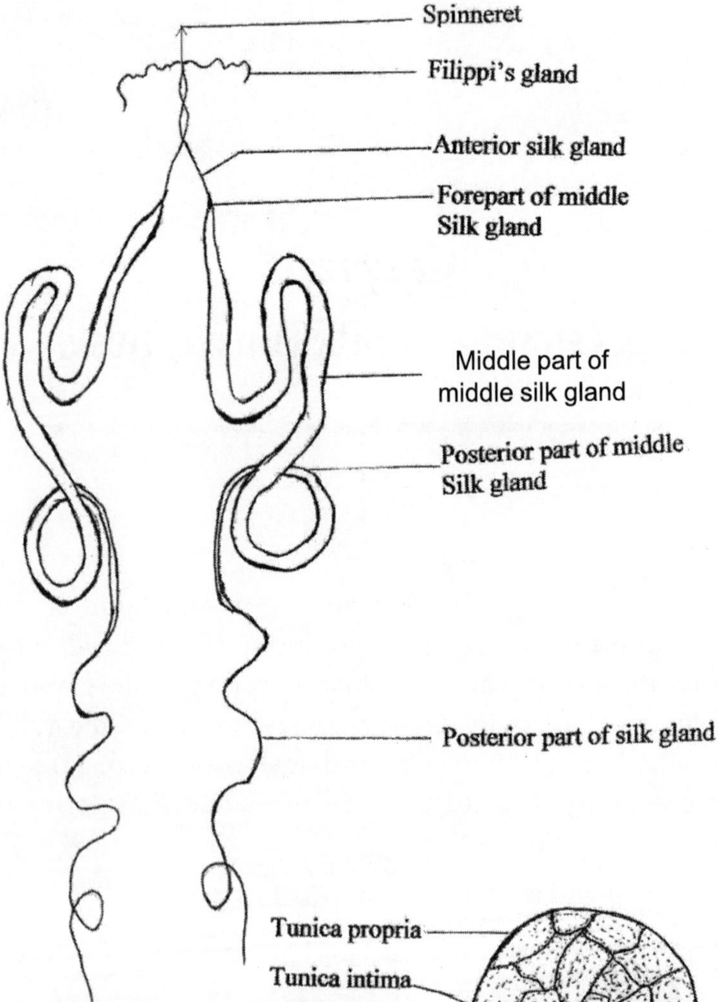

Silk Gland of Silkworm

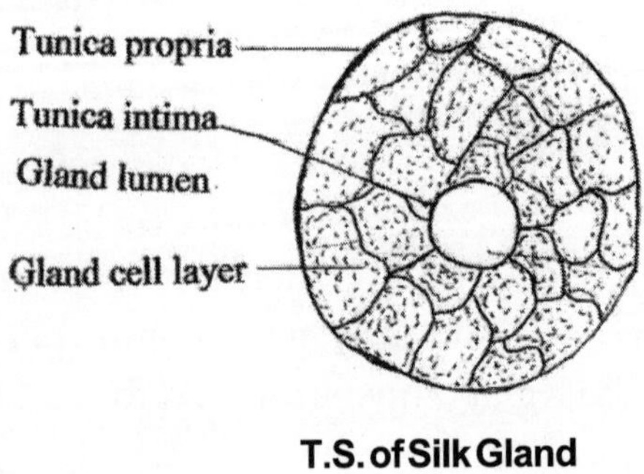

T.S. of Silk Gland

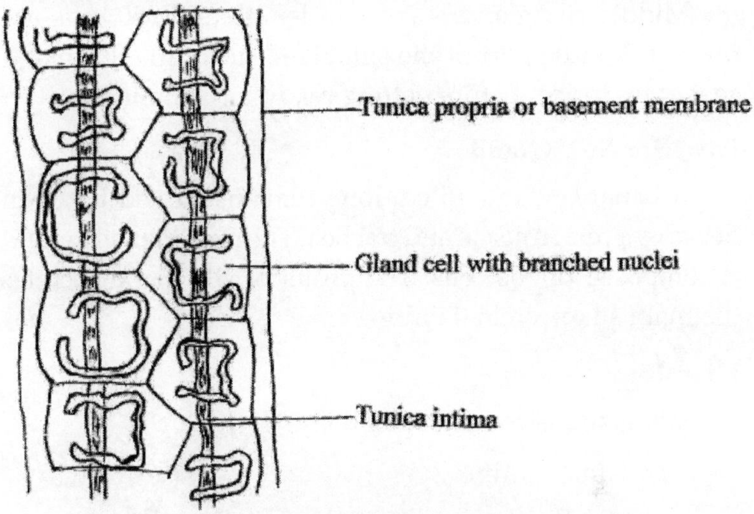

L.S. of Silk Gland

Anterior part of the silk gland has no secretory function. It serves as a passage of silk while the Fillipi's gland secretes the mucus which covers the silk filament.

Middle Part

It is again sub-divided into three parts

Fore Part

Narrow anteriorly but gradually widen posteriorly. It secretes sericin III

Mid Part

It is thick slender lies between pre and hind parts of middle silk gland. It secretes sericin II

Hind Part

It gradually narrows towards posterior end and ends in posterior silk gland. It secretes sericin I

Middle silk gland is 7 cm in length and 350 cells are present. Middle part of the silk gland stores the fibrin and coated by sericin before getting ready for spinning.

Posterior Silk Gland

It is narrow and 15 cm long remains curved between dermo-visceral muscle and trachea. The posterior silk gland is composed of 500 cells. The posterior silk gland secretes the main silk protein–Fibroin.

Histology

Silk gland is composed of 3 layers

1) Tunica intima
2) Middle gland cell with branched nuclei and
3) Outer tunica propria

At the centre there is a gland lumen.

BIO SYNTHESIS OF SILK

Silk is a protein. Main protein in silk is Fibroin and sericin

Sericin

It is the outer layer of fibroin, which is gummy in nature. It is rich in sericin, glycine and Aspartic acid.

Fibroin

It has two sub-units

1) ***Larger sub unit***: its molecular weight is 3, 50,000 and it is rich in alanine, glycine, sericin and tyrosine.
2) ***Smaller sub unit***: its molecular weight is 25,000 and it is rich in aspartic acid and glycine

As silk is a protein, so the biosynthesis of silk follows the same principle which is observed in general protein synthesis. Only the sequence of amino acids in the polypeptide chain is changed.

The proteins are long chain polymers of amino acids. These amino acids are linked by a covalent bond called a peptide bond between the amino (NH_2) group of one amino acid and carboxyl (COOH) group of another. In each polypeptide chain a free carboxyl group always occurs on one end and a free amine group to the other (the C-terminal and N-terminal ends).

The following steps are involved in protein synthesis. The central dogma in protein synthesis has been described as:

$$\text{DNA} \xrightarrow{\text{Replication}} \text{DNA} \xrightarrow{\text{Transcription}} \text{RNA} \xrightarrow{\text{Translation}} \text{Protein}$$

$$\circlearrowleft \text{DNA} \rightleftarrows \text{RNA} \longrightarrow \text{Protein}$$

A. Replication or Duplication

In this process each DNA strand act as templates and replicate its own replica in the presence of deoxy-ribo nucleotide and Mg. The base sequence which is present in the DNA molecule is utilized to protein synthesis.

B. Transcription

The genetic information transcribed in the base sequence of the DNA is copied in the form of RNA. Newly synthesized RNA undergoes post transcriptional modification and produce active RNA molecules. These are mRNA (messenger RNA). It supplies the information as codon in the protein synthesis.

tRNA

It acts as the translator and it has anticodon.

rRNA

It unites with ribosomes and form 80s ribosomal subunit (S=Svedberg Unit). (The Svedberg unit(s) is a measure of sedimentation velocity of particles in the ultra centrifuge). 80s ribosomal unit has two sub units a) 40s sub unit: It is smaller and has 1 RNA molecule and 32 proteins.

60s sub unit: It has 3 RNA molecule and 45 proteins.

C. Translation

It has again divided into 3 stages:

(*i*) Initiation

(*ii*) Elongation

(*iii*) Termination

Initiation

The following steps are observed in this stage

a) Activation of amino acid

b) Formation of amino acyl tRNA

c) Binding of amino acyl tRNA to ribosomal complex

Translation

Like the synthesis of general protein, the translation of fibroin from mRNA also needs three stages

1) Promote reaction
2) Elongation reaction and
3) Termination

The synthesis of fibroin begins from methionine, that is the methionine codes at the fibroin mRNA is the beginning site, to synthesize the fibroin protein. In fact the N-end of the fibroin molecules is not methionine. This is because through process after transcription,surplus peptide at the N-end chain containing methionine is excised.

During the process of fibroin synthesis, there are three elongation factors: EF1a, EF1b and EF2. They participate in the elongation reaction of peptide chain. They play a role in peptidyl transfer reaction and translocation reaction. When the elongation peptide chain arrives at the termination codes (UAA, UAG or UGA), the synthesis of polypeptide is ended and the polypeptide is released from the ribosome. At this time, releasing factors are needed in order to terminate the reaction.

Chapter 6
Nervous System and Sensory Organ

The Nervous is the leading physiological systems, it monitors both internal and external environment of the body through the sensillae and sensory nerves, integrates all information in the central nervous system, signals to control all the activities of the various parts of the body.

Nervous System

Nervous System in the Larval Stage

The nervous system in the larval stage includes the central nervous system, peripheral sensory nervous system and visceral or sympathetic nervous system.

The Central Nervous System

It is located on the median ventral surface beneath the integument, and it is a chain and key part of the nervous system consists of the paired ganglia connected by connectives, but the members of the paired ganglia are usually so closely fused that they appear to be a single ganglion. The ganglia include the ganglia in the head, thorax and abdomen.

There are two pairs of ganglia in the head. One is the brain, the first pair of ganglia, it is located on the dorsal side of the pharynx and composed of two semi-globular closely fused ganglia. It is the centre for the connection and regulation of all the nervous actions. It can secrete brain hormone (BH), and control the secretion of other hormone. In addition the suboesophgeal commissure which arises from the brain and surrounds there are two circumoesophageal connectives which connect the brain with the sub-oesophgeal ganglia. The sub-oesophgeal ganglia are the second pair of the ganglia in the head; they are closely fused and located on the ventral side of the pharynx and appear to be triangular. From the suboesophageal gangla two thick short connectives arises to connect with the ganglia in the thorax. This ganglia functions mainly as a controller for the action of the various parts of the mouth, and can secrete diapause hormone (DH).

There are three pairs of ganglia in the thorax. They are located on the ventral sides of the prothorax, mesothorax and metathorax and called the third, fourth and fifth pair of ganglia, respectively. Each paired ganglia are fused so closely that they appear to be a five angular ganglion. There are two thick short connectives between the anterior and posterior ganglia. These ganglia innervate the muscles of the corresponding segments. The fourth paired ganglia innervate all the spiracles on the body.

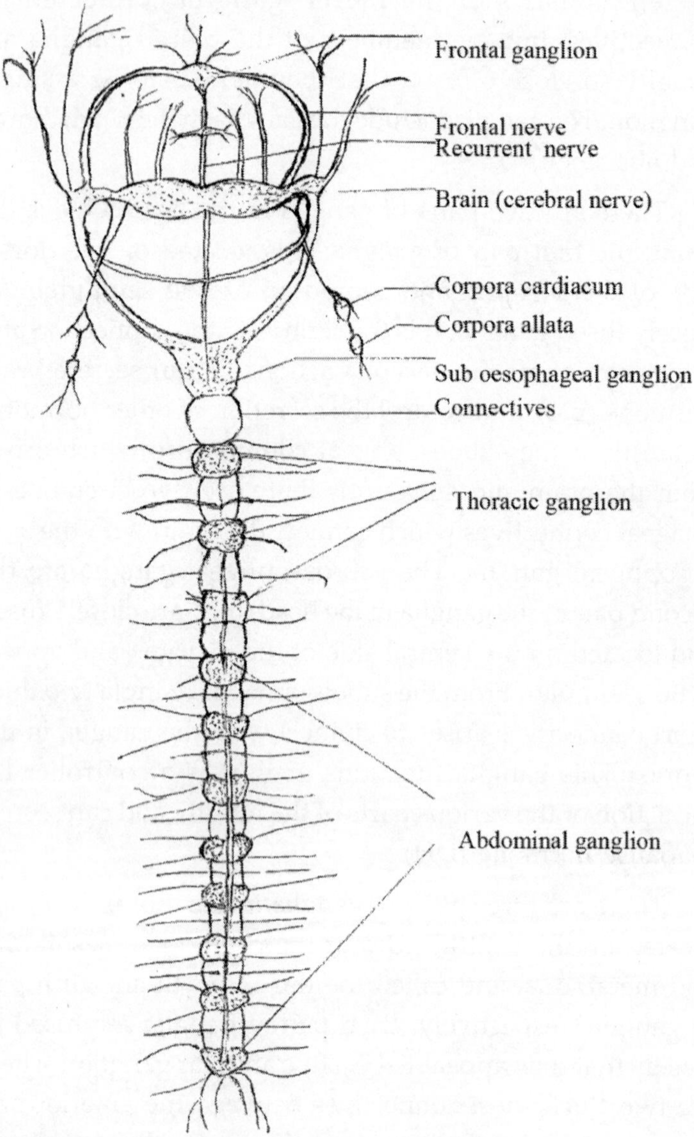

Lateral Nervous System

There are eight pairs of ganglia in the abdomen. Each paired ganglia are fused and appear to be a hexangular ganglion. The eight pairs of ganglia are called the sixth, seventh,eighth,ninth,tenth eleventh,twelfth and thirteenth pair of ganglia, and located on the ventral side of the anterior seven abdominal segments, respectively. Among them the last two pairs are so closely fused that they form a composite ganglion.

Therefore the ganglia on the seventh abdominal segment are composed of the last two pairs of ganglia. A connective connects the anterior ganglion with posterior ganglion. The connective in fact is paired connectives, but they are fused so closely that they appear to be one connective. In addition to control the corresponding segmental respiratory movement, the abdominal ganglia innervate the dorsal vessel, and control the spinning, the abdominal muscular movement,fecal cxcretion and reproductive functions.

The Peripheral Sensory Nervous System

The peripheral nerves are branches of the central nerves and distributed beneath the integument on the peripheral tissues and organs. There is a front nerve in front of the brain, which connects brain with the frontal ganglion. On the lateral sides of the brain arise three pairs of nerves, they innervate labrum, antennae and ocelli and are called labral nerves, antennal nerves and visual nerves.

Three pairs of nerves arise from the anterior edge of the sub-oesophageal ganglia, they innervate the mandibles, maxillae and labium and called mandibular nerves, maxillary nerves and labial nerves respectively. Two nerves arise from the posterior lateral edges of the sub-oesophageal ganglia, they prolong backwards to the prothorax and rest on the lateral sides of the dorsal vessel.

In thorax and abdomen, two pairs of nerves arise from both left and right side of each ganglion from the front and hind edges, but composite ganglia bear no such nerves. The anterior pairs of the nerves are longer, they prolong toward side and rest on the dorsal muscles and dorsal integument at the middle of the nerves, and they frequently arise short fine branches which rest on the inner layers of the ventral muscles. The posterior pairs of the nerves are short and thin, rest on the outer layers of the ventral muscles and in the thoracic and the abdominal legs. The composite ganglia arise three pairs of nerves, the middle pairs equal to the connective of other ganglia, while the outer two pairs equal to peripheral nerves of other ganglia, inner pair of which rest on the rectum around anus and in the caudal legs, etc.

The Visceral or Sympathetic Nerves

The visceral or sympathetic nerves are also the branches of the central nerve, but they are distributed mainly on the visceral surfaces. A short, fine frontal nerve arise anteriorly from the frontal ganglia whose branches innervating beneath the dorsal wall of both the buccal cavity and pharynx. A median long recurrent nerve, which arises posteriorly from the frontal ganglia prolongs backwards along the ventral side of the dorsal vessel, and is divided into two nerves near the oesophagus, one of them innervates the oesophagus and the anterior mid gut, the other innervates the dorsal vessel.The recurrent nerve controls the movement of the pharynx. From the bilateral posterior edge of the brain arise a nerve, which immediately connected to the corpora cardiaca then connected to the corpora allata through the lateral oesophageal sympathetic nerve. In addition from the brain many fine nerves arise to control the mandibular dialator muscle, pharynx, dorsal vessel, salivary gland and oesophagus etc.

The sympathetic nerve include the nerves which arises from the posterior edge of each ganglion, and prolongs backwards to slightly front of the coming ganglion, where the nerve branches bilaterally in to two nerves. These two nerves go by the central longitudinal muscles, rest on the closing apparatus of the spiracle in the coming segment, and connect to the nerve arisen from the anterior edge of the ganglion in the coming segment, their connections rest on the dorsal longitudinal muscles. The sympathetic nerves arise from the composite ganglion control the rectum and the reproductive organs. The sympathetic nerves intercross the trachea to form the sympathetic nervous webs.

Nervous System in Adult

The nervous system in adult also consists the central nervous system,peripheral nervous system and sympathetic nervous system too. But in adult, the brain and sub-oesophageal ganglion are much bigger and the brain is much well-developed. Because the bilateral connectives of the pharynx have shortened, the brain and the sub-oesophgeal ganglion are very near, the distances between the thoracic ganglia are very short,while the las two thoracic ganglia are so closely fused that they appear to be a spoon-shaped ganglion. In abdomen, only 4 pairs of ganglia can be recognized because of the combination during metamorphosis, furthermore the connectives are getting shorter and thicker.

The peripheral and sympathetic nerves in adult are basically the same as that in the larval stage, but no sympathetic nervous webs are observed in adult.

Histological Structure of Nervous System

The nervous system is composed of numerous neurons and the nerve cells. The neuron is not only a structural unit but also a physiological functional unit. The neuron receives the stimuli from both internal and external environment as well as transfers the impulse caused by the stimuli.

Neuron and its Type

A neuron comprises a cell body (perikaryon) and its processes. The processes form the nerve fibre. The cell body appears to be globular, its nucleus is located at the centre of nerve cell, in which, plasma, mitochondria and other organelles are filled with. There are two kinds of processes.One is the dendrites which are short, branched outgrowths of the perikaryon. It branches repeatedly to increase its functional surface area for receiving the impulse. Its function is to receive the impulses come from other neuron and transfer them to the perikaryon. The other process is the axon, which is an extremely elongate cytoplasmic process. This process often has a collateral branch at its base. Both the axon and its collateral branch often branched distally to form the terminal arborization, which are short fine branches, axon's function is to transfer the impulses coming from its perikaryon to other nerve cells or to the effector organs.

According to the number of process, the neurons are divided into three types: the monopolar neurone, bipolar neurone and multipolar neurone. In the monopolar neuron, its perikaryon has only one process which has a collateral branch on the process base. In the bipolar neuron,the perikaryon has an axon and a short process, which is called dendrite while the perikaryon in the multipolar neuron has axon and dendrites as well.

According to the functions the neurons are divided into three types: the sensory neuron, motor neuron and internuncial (association) neuron. The sensory neuron receives the stimuli through the sensillae and transfer impulses caused by them to the neuron in central nervous system to effector organs. While the association neurons which play associating role between the sensory and motor neurons.

The Structure of the Ganglion and Nerve

The ganglion in fact is an aggregation of numerous neurons and their nerve fibres. The aggregation includes mainly the perikarya and their axons, collateral branches and their terminal arborizations of the motor neurons and internuncial neurons as well as the axons and their terminal arborizations of the sensory nerves.

In a ganglion there is a central neuropile which comprises a mass of nerve fibres to form a composite structure of nerve network. The central neuropile is the centre of the synaptic association for the central nerves and a key part of the association and the regulation for the nervous system. Surrounding the neuropile is the perineuropile tissue in which the perikarya of motor and association neurons are normally found adjacent to the neuropile. The perikarya of the motor neurons are larger and located slightly in the inner site of the tissue, their cytoplasm being uniform. While the perikarya of the association neurons are small and located slightly in the outer side of the tissue.with little cytoplasm. The perineuropile tissue is enclosed by the nerve sheath which is divided in to two layers. The outer layer is the acellular nerve lamella, a protective sheath which contains collagen fibrils and neutral mucopolysaccharide. The lamella has not only flexibility but also resistibility to bulge and regulate the ionic flow by the lamella. The inner layer of the nerve sheath is composed of galial cells and called the perineurium. These galial cells are loaded with mitochondria, glycogen oxidase phosphatide and cerebroside and serve as a screen of ionic selection. In addition, some galial cells in the outer of the layer surround the perikarya of the neurons some galial cells in the inner of the layer rest on the surface of the neuropile and extend finger like extensions of the cytoplasm in to the neuropile.

Nerve comprises the axial cord and nerve sheath. The axial cord is the centre of the nerve, and a single nerve fiber. Single axial cords are associated by the galial cells to form bundles of the axial cords. Each nerve fibre can transmit the nerve impulses and is enclosed by a thin axial membrane, in which there are axoplasm, loaded with mitochondria. The nerve sheath is the outer covering, its structure is similar to that in the ganglia.

Sensory Organs and their Functions

The sensory organs receive the stimuli from both internal and external environment, and change the stimuli, which exist frequently in form of energy, into the nervous impulses, and transmit them to the central nervous system. The sensory organs also receive and transmit the message coming from the other intraspecific individual. The sensory

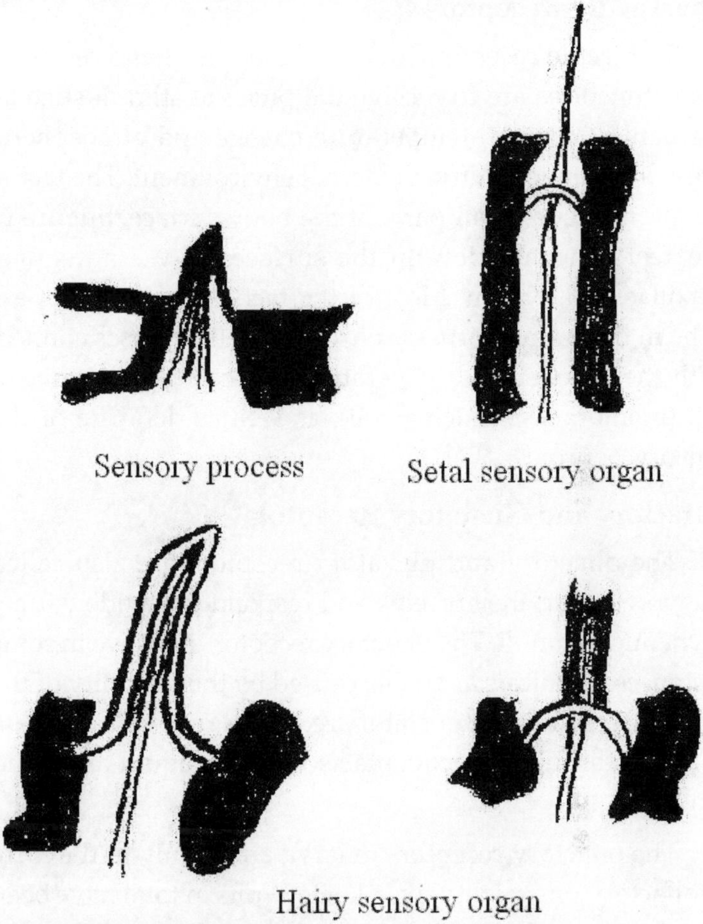

Different Types of Sensory Organs in Integument

organs commonly are classified by the nature of the stimuli that activate them. However, here the sensory organs are classified by means that the sensory organs receive the stimuli they are tactile receptors, olfactory receptors, gustatory receptors, visual organs and so on.

The Tactile Receptors

The tactile receptors are also called mechano receptor. Their functions are to receive the physical stimuli such as mechanical contact, temperature change and atmospheric pressure change etc. from external environment. The tactile receptors occur on all parts of the body surface, but are in greatest concentration on the surfaces of the antennae, maxillae and labium. Most of the tactile receptors appear to be in the simple form of sensory hair, their bases connect with the tricogen cell (generative hair cell) and tormogen cell (membrane-poducing cell), as well as dendrite of the sensory neuron.

Olfactory and Gustatory Receptors

The olfactory and gustatory receptors are also called chemoreceptor, they receive and respond to a wide variety of chemical stimuli. The olfactory receptors perceive in some distances chemicals as smells caused by the volatility of the substances, while the gustatory receptors receive the stimuli as tastes through direct contacts with the liquid and solid materials.

The olfactory receptors in larva are mainly in the form of olfactory pits, ninety five olfactory pits in total have been found, among them, twenty two are on the head capsule, thirty eight are on the head appendages and remaining thirty five on the trunk and its appendages. The olfactory pit appears frequently to be a discoid membrane with black scleritized outer edge and micropores in central area. Three to four conic processes on the second segment and three conic processes on the third segment of each antenna and eight conic processes on the terminus of the each maxipalpus in larvae are the examples of olfactory receptors.

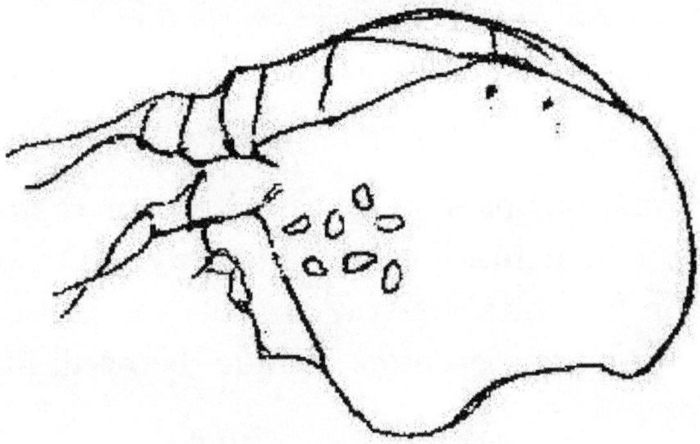

Olfactory Pore in Head

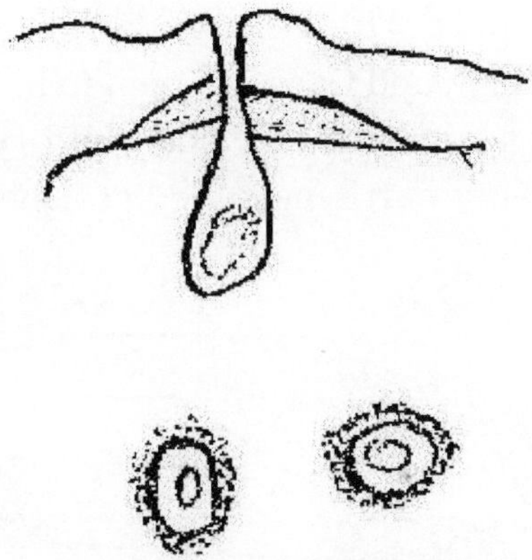

Structure of Olfactory Pore

The trichoform processes on the male antennae are also the olfactory receptors, and they sense the sex pheromone, and facilitate location of a mate.

The gustatory receptors are common on mouth parts. The three processes on the inner labial surface, especially the two segmented processes on the maxillary tubercles play important role in selection of the food. Each gustatory receptor connects to particular sensory neurons, which repond to special substance. Some receptors respond only to saccharide, while others respond to salt and water.

Visual Organs

The visual organs are also called photoreceptor, they perceive the structure of light and wave, transmit the information to central nervous system and cause visual response. The photoreceptors include the ocelli in larvae and compound eyes in adult.

The ocelli and compound eyes are similar in structure; they are composed of two parts: light gathering part and lightsensory part. The light gathering part serves as gathering the light and transmitting it to light sensory part, and comprises cornea glassy body and crystalline cone, while the sensory part senses the light energy and causes nerve impulses.

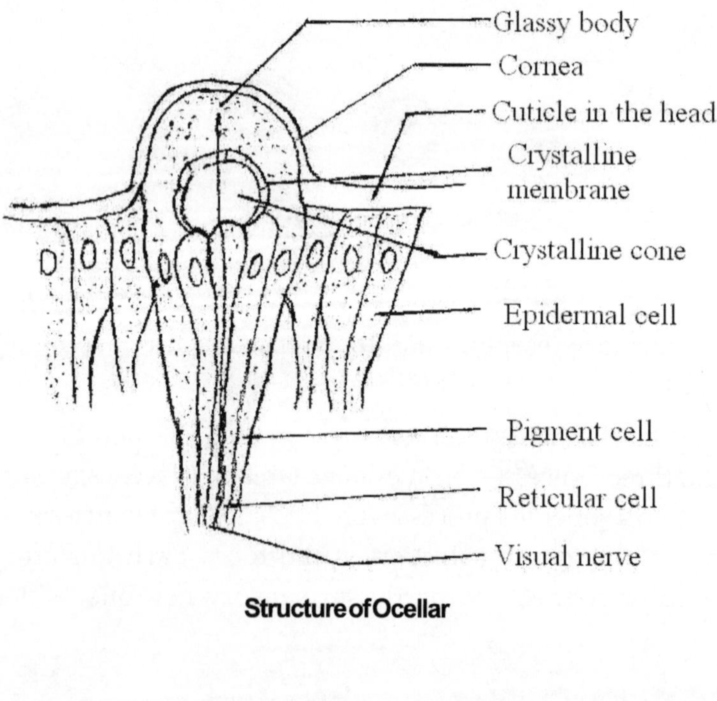

Structure of Ocellar

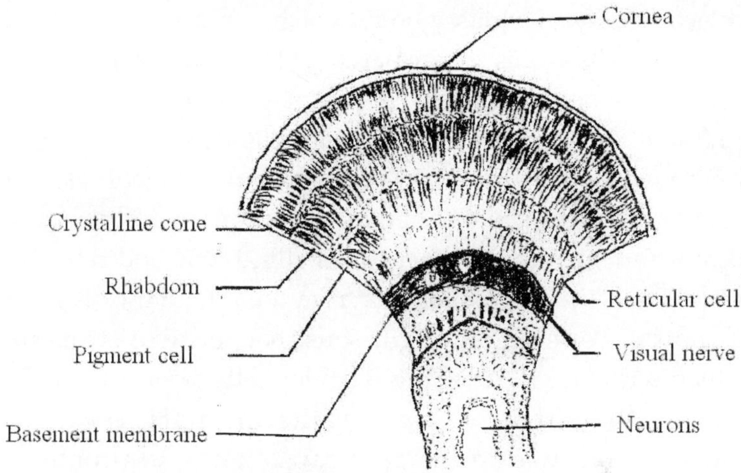

Section of a Compound Eye Basement

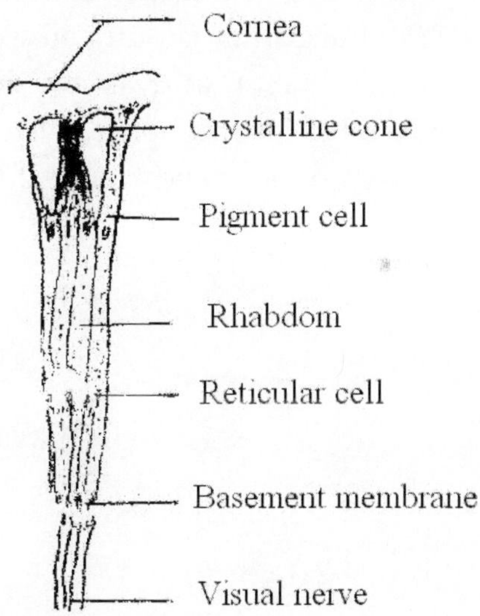

Enlarged Portion of a Compound Eye

The cornea of the ocelli is transparent, colourless, modified cuticle, and convex to appear as a lens. Beneath the cornea lies the glassy body which was formed by three blended cells and is a transparent colourless plasmic layer. The glassy body connects crystalline cone. There are three pigment-containing long reticular cells beneath the cone. There were two corneal cells at first under the cornea, but finally they reduce or change into pigment cells, move down and surround both the crystalline cone and reticular cells. The dendritic extensions from the reticular cells from a central, coumpound, light-sensitive rod like structure called a rhabdom. After the reticular cells receive the light transmitted from the light gathering part, resultant chemical reaction in the cells cause nerve to produced nervous impulses and to visualize.

Each compound eye in adult is composed of 3000 ommatidia. Each ommatidium is outlined externally as a hexagonal cuticular facet and is an independent photoreceptor, which has its own optical system and visual cells, and has its own optic nerves connecting the visual centre.

Chapter 7
Digestive System of Silkworm

In the silkworm larva the digestive system is more or less straight tube. It starts from the mouth and ends into anus. It is divided in into three main parts: the fore gut or stomodeum, mid–gut or mesenteron and hind gut or proctodeum.

The fore gut of the stomodeum consists of (1) oral aperature (2) Mouth cavity (3) Pharynx and (4) Oesophagus. The oral aperature opens into the mouth cavity which is followed by a narrow pharynx and oesophagus. The oesophagus is narrow at the anterior end and gradually widens towards the posterior end. There is a cardiac or stomodeum valve at the end of the fore gut that retains the chewed mulberry leaf bits in the oesophagus for some time and also prevents the regurgitation of food from midgut to fore gut. The midgut is long, wide cylindrical tube narrow at the posterior end. Digestion and assimilation of food take

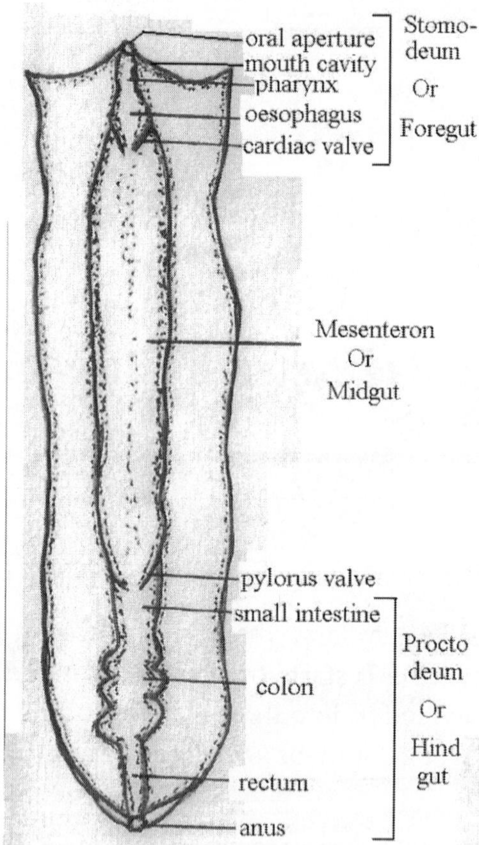

Digestive System of Silkworm Larva

place mainly in the mid gut. The digestive fluid is secreted principally from the goblet cells at the mid gut epithelium and the cylindrical cells absorb the digested food. The fore and the hind guts have a chitinous lining, but an inner layer of peritrophic membrane in the midgut generally protects the mid-gut epithelium from mechanical damage due to food particles. The hind gut consists of the (1) small intestine (2) colon and (3) rectum. There is a pylorus valve near the anterior end of the small intestine which guards and regulates the passage of digested food from the midgut to anterior hindgut.

The hindgut is a passage for the absorption of a large portion of food moisture and elimination of digested food. In the anterior-posterior direction, the rectum has six muscles for pressing the excrements. The faecal matter is pressed in the rectum and expelled from the anus as faecal pellets bearing hexagonal marks.

According to the function,the alimentary canal of silkworm larva may be divided into four regions such as (1) Region of reception (2) place of conduction and storage (3) Region of digestion and assimilation (4) Region of conduction and formation of faeces.

The oral aperture, mouth cavity and pharynx are the regions functioning for reception of the food material while oesophagus acts as the store house of food material. The mid intestine acts as the region of digestion and absorption while the hind intestine acts as the region of absorption, conduction and formation of faeces.

In the adult the proboscis is vestigial as well as nonfunctional and the adult does not feed. Hence, the adult digestive system appears degenerated and less prominent than the larval digestive system. The alimentary tract is formed of a long oesophagus of narrow caliber with a diverticulum (sucking stomach) of the fore intestine appears to be a main receptacle for the air that is swallowed and distends the body after metamorphosis. Since, the adult takes no food, this may be its main function. A couple of salivary glands open in to the oesophagus. Six Malpighian tubes are present, three arising from each side of the hind gut by a common tube. The hind intestine enlarges into a rectal sac and ends into the anus by a short tube.

Ingestion

Ingestion is a process by which food material is being taken inside the alimentary canal. The silkworm cut the mulberry leaves by the mandibles right and left and taken in to the mouth cavity. At the time of cutting mulberry leaves mandibles are carried from side but the edges of mandibles turn somewhat inwards, both edges going up and down alternately as doors open and close.

Factors Influencing the Ingestion

1) Physical and chemical properties of the leaves.
2) Presence of nutrients in balanced proportion.
3) The rate at which the alimentary canal is empty.

The amount of ingestion is calculated by the following formula.

$$\text{Amount of ingestion} = \text{Amount supplied} - \text{Amount not eaten}$$

$$\text{Ratio of ingestion} = \frac{\text{Ingestion amount}}{\text{Supplied amount}} \quad 100$$

Digestion

It is the process of breakdown of complex foodstuffs (materials) in to simpler compound which can be incorporated into metabolism. In other words it may be defined that it is the procedure of changing insoluble substances into soluble substances and also the impermeable substances into permeable substances.

The nutrient content of the mulberry leaves are not soluble in general and covered by the cell walls, which consists of cellulose, accordingly they are not utilize by the silkworm directly. In order to take out the nutrient from cells of mulberry leaves. Silkworm must change the insoluble matter into the soluble one. Among the several divisions of the alimentary canal of silkworm, the mid-intestine has the main important role for digestion and absorbtion. The digestive juice is consisted of saliva and gastric juice. The saliva is a weak alkaline solution containing an enzyme called amylase. Gastric juice or gastric fluid is a strong alkaline solution. The pH of gastric juice is 9.2–10.3. The principal enzymes contained in the gastric juice are tyrosinase (a protease), lipase (fat decomposing enzymes), amylase, maltase and glycogenase (carbohydrate decomposing enzymes), oxidase, peroxidase, catalase and tyrosinase (oxiding enzymes). Amylase decomposes starch into dextrin and maltose. Maltose decomposes maltose into dextrose. Oxidase has a free oxygen and performs oxidation. Peroxidase performs oxidation in co-operation with hydrogen peroxide.

Invertase exists in the cells of the intestine and decomposes sucrose in to glucose and fructose. Glactase exists also in cells of the intestine and hydrolyzes galactan into glucose and fructose (or galactose). These two enzymes are not found in the gastric juice.

Absorption and Assimilation

Absorption

The passage of food through the lining of the digestive tract in to the blood directly or indirectly through the lymph is called absorption. Among the components of mulberry leaves, some elements, such as water glucose soluble salt etc. are absorbed directly by the cells of the mid-intestine, but protein and other molecule of big size cannot permeate the cell wall of the mid intestine. So they should be decomposed into simple forms and absorbed.

Absorption of Protein

Protein is decompose into amino acids and peptone, which are now permeable to the cell wall of the mid-intestine. This decomposition is catalysed by the enzyme known as trypsin. It is absorbed directly by the haemolymph. So, after a feeding the amount of amino acids in the haemolymph increases.

Similarly, fat is decomposed into fatty acids and glycerine by lipase.

The main components of the carbohydrate of mulberry leaves are starch, sucrose and glucose, among these glucose is absorbed directly, but starch is decomposed by amylase and maltase and at last it turns into glucose. Sucrose is decomposed by saccharase in to glucose and fructose and then absorbed.

Assimilation

There is difference in the protein of mulberry leaves and the protein of silkworm body. The proteins of mulberry leaves is decompose by an enzyme into amino acids and peptone and absorbed by the cells of the mid-intestine and ultimately carried to the respective organ through haemolymph. In the respective organ the amino acids and peptones are again synthesized in to animal proteins. This process is called assimilation. Others matter also assimilated in the silkworm body, such as fat is synthesized from the fatty acids and glycerol. Besides this fat is also synthesized from carbohydrates. A greater part of mulberry carbohydrates synthesized into fats in the silkworm body.

Undigested Substances

The greater part of the silkworm excrement (Faeces) is the undigest compound of the mulberry leaf and water. The water content of the silkworm excrement increases with advance of the age of larvae. It is 30 per cent in first instar 55 per cent in 4th instar,64 per cent in fifth instar and 74 per cent in matured larva.

Consumption, Growth and Utilization Indices of Food

To understand the utilization of food by an organism requires the answers of the following questions.

1) At what rate is the food is eaten?
2) How much of the eaten food is actually digested?
3) What part of the food is incorporated as body substance

These questions cannot be answered without (1) a reliable and accurate method for determining the weight of the food ingested,(2) the weight of the faeces which correspond to the ingested food and (3) the weight gained by the insect during the experiment (here during the instar).

These three measurements are basic and necessary for the calculation of (1) the rate of feeding (2) the digestibility and (3) the efficiency of conversion of food to body substance.

The indices of the digestive system mentioned below can be calculated on the basis of fresh or dry weight but it should be noted that fresh and dry weight indices are not comparable since the percentage dry matter of food, faeces and insect are likely to differ.

Consumption Index or C.I.

The rate of feeding (intake) relative to the mean weight of the animal in a definite time can be expressed as consumption index or C.I.

$$C.I. = F/TA$$

where,

F: Wt. of leaves consumed.

T: Duration of feeding period (days).

A: Average wt. of the larvae during the period.

Growth Rate

It is a measurement that explains the rate at which the digested matter is available to the silkworm (animal) during the experimental period (here instar).

$$G.R. = \frac{\text{Wt.gained by the larva}}{\text{Duration of feeding period (in days)} \quad \text{Av. (Mean) Wt. of the larva during the feeding period}}$$

Efficiency of Conversion of Ingested Food (E.C.I.)

This measurement indicates the overall efficiency of the animal to utilize the food for growth. This value is dependent upon the digestibility of food, its nutritional value and the level of nutrient intake. The percentage efficiency of conversion of food into body material was calculated by dividing the weight gained by the larvae by the weight of food consumed. This ratio can also be obtained by dividing one rate by another.

$$\text{E.C.I.} = \frac{\text{Weight gained by the animal}}{\text{Weight of the food ingested}} \quad 100$$

Or

$$\text{E.C.I.} = \frac{\text{Growth rate (G.R.)}}{\text{Consumption Index (C.I.)}}$$

Approximate digestibility (A.D.)

This is represented by the equation.

$$\text{A.D.} = \frac{\text{Wt. of food ingested} - \text{Wt. of faeces}}{\text{Wt. of food ingested}} \quad 100$$

The weight of food consumed = wt. of food introduced – (minus) the weight of leftover food. It (A.D.) is also termed as (1) Co-efficient of utilization (2) Digestion co-efficient (3) Co-efficient of digestibility,(4) Degree of absorption for expression for expressing the digestibility of food material. Waldbauer (1968) pointed out that this measure is misleading and should be referred to as approximate digestibility. This is true because unlike mammals in insects the value obtained by subtracting weight of the faeces from the weight of the food digested. Waldbauer(1968) has discussed this aspect in great detail and noted that correction of the urine content of the faeces would result into a value which could be termed as co-efficient of apparent digestibility (C.A.D.). Such correction would result into a meaningful value.

Efficiency of Conversion of Digested Food (E.C.D.)

The measurement of the efficiency with which the

digested food is converted into body substance (E.C.D.) is calculated as:

$$\text{E.C.D.} = \frac{\text{Weight gained by the animal}}{\text{Weight of food digested}} \quad 100$$

The amount of food digested can be calculated by subtracting the weight of faeces from the weight of food ingested. The results can be computed on fresh or dry weight basis and the final value can be expressed in per cent. E.C.D. values will vary with the level of intake and nutritional value of food materials and is independent of digestibility. Some of the authors defined this index as co-efficient of growth (QgD) but Waldbauer defined this index as efficiency of conversion of digested food to body material. This terminology appears to be more appropriate because the amount of food utilized is actually the amount of food digested.

The Utilization of Dietary Constituents

The utilization of nutrient constituents of the natural and chemically defined food is generally done by the determining the amount of a nutrient required per weight or volume of diet. Minimum or optimum requirements expressed in these terms say nothing about the absolute quantities required, but define only the relationships between the requirements for particular nutrients. This sort of information facilitates the formation of diets, but convey a minimum of biological meaning pertinent to questions of metabolism, ecological relationships or the adaption of insects to their natural foods.The absolute requirements have been neglected because their determination requires to the laborious task of measuring food intake.

The co-efficient of apparent digestibility for the constituents "X" of the food is usually expressed as a percentage and is calculated as:

$$\text{C.A.D.} = \frac{\text{Amount of "X" in food ingested} - \text{Amount of "X" in faeces}}{\text{Amount of "X" in food ingested}} \quad 100$$

The efficiency with which the digestible portion of the constituent "X" is converted to body substance is calculated as:

$$\text{E.C.D.} = \frac{\text{Amount of "X" in body}}{\text{Amount of "X" in food ingested} - \text{Amount of "X" in faeces}} \quad 100$$

The efficiency of conversion of ingested constituent "X" to body substance is calculated as:

$$\text{E.C.I. (X)} = \frac{\text{Amount of "X" in body}}{\text{Amount of "X" in food ingested}} \quad 100$$

Utilization of Energy

The study of the bio-energetics explains the follow of energy from one trophic level to another trophic level. The parameters for measuring the energy are (1) Co-efficient of metabolized energy (C.M.E.), the efficiency of storage of ingested energy (E.S.I. (E) and the efficiency of storage metabolized energy (E.S.M. (E).

C.M.E. is calculated as:

$$\text{C.M.E.} = \frac{\text{Gross energy in food consumed} - \text{Gross energy in faeces}}{\text{Gross energy in food consumed}} \quad 100$$

For obtaining a meaningful measurement on the energy utilization by insects, it is essential to correct the gross energy in feces by subtracting the gross energy of urine content. Similar steps should be taken for correcting the gross energy of body by substracting the gross energy of uric acid stored in the fat body. Hiratsuka (1920) studied in detailed the relationship of energy intake and its expenditure for maintaining life of ***Bombyx mori.*** The C.M.E., E.S.I. (E) and E.S.M. (E) values varied from

C.M.E. = 72-82

E.S.I. (E) = 28-32

E.S.M. (E) = 61-68 per cent respectively.

The last instar stores a large amount of metabolized energy for the non-feeding pupal stage. Our knowledge on the follow of matter through insects is far from perfection because measurement of gross energy of food consumed, gross energy in the feces and body of the animal requires laborious work and also in many cases it is technically difficult. In recent years insect physiologist and ecologists are focusing their efforts in working out energy relationship and follow of matter through insects.

The gross energy (or heat of combustion) of a substance is usually measured with a bomb colorimeter and is defined as the energy liberated as heat when the substance is completely oxidized to CO_2 and H_2O.

Metabolizable energy has been defined as the gross energy in the food eaten less the gross energy in the faeces and urine. This is the energy available for the production of heat, body substance or work.Hiratsuka (1920) subtracted the gross energy in the faeces (urine included) from the gross energy of the food eaten by the ***Bombyx*** and called this value the physiologically available energy. He made no mention of the possibility that insects may produce digestive gases.

The efficiency of the storage of ingested energy is calculated as

$$\text{E.S.I. (E)} = \frac{\text{Gross energy stored in body}}{\text{Gross energy in food eaten}} \quad 100$$

and efficiency of storage metabolized energy E.S.M. (E) is calculated as follows:

$$\text{E.S.M. (E)} = \frac{\text{Gross energy stored in body}}{\text{Gross energy in food eaten} - \text{Gross energy in faeces}} \quad 100$$

Hiratsukas (1920) study of ***Bombyx*** remains to this day the most complete study of the intake and expenditure of energy by an insect.

Table 1: Comparision of the Utilization of Dry Matter and Energy by *Bombyx mori*

To End of the 5th Instar		*To Newly Emerged Adult*	
Dry Matter	*Energy*	*Dry Matter*	*Energy*
A.D. 37	C.M.E. 42	A.D. 37	C.M.E. 42
E.C.I. 23	E.S.I. (E)28	E.C.I. 8	E.S.I. (E) 12
E.C.D. 62	E.S.M. (E) 67	E.C.D. 22	E.S.M. (E) 67

Table 1 shows that the utilization of energy is higher than the utilization of dry matter but that the two are roughly comparable. Hiratsuka (1920) also found that the consumed portion of the leaves had a somewhat greater heat value than the total leaves supplied to the larvae.

Table 2: The Utilization of Energy by the Larval Instars of *Bombyx mori*

Inster	*C.M.E.*	*E.S.I. (E)*	*E.S.M. (E)*
I	52	32	61
II	49	29	60
III	42	28	66
IV	44	29	66
V	42	28	68

Table 2 summarizes the utilization of energy by the five larval instars of ***Bombyx mori***. The C.M.E. and the E.S.I. (E) vary with age much as do the corresponding A.D. and E.C.I.of dry matter (Table 3), declining in the early instars and then tending to remain more or less constant.

Table 3: The A.D. and E.C.I. of *Bombyx mori* from Instar to instar. Hiratsuka (1920).

Index	*Instar*				
	I	*II*	*III*	*IV*	*V*
A.D.	47	44	38	38	37
E.C.I.	31	27	25	25	23
E.C.D.	66	61	66	65	61

The E.M.S. (E), however, increases with age, reaching its peak in the 5th instar, while the corresponding E.C.D. of dry matter tends to remain relatively constant from instar to instar.

The large amount of energy stored during the 5th instar is of course, needed to support the nonfeeding pupa and adult.

Effect of Environmental Factors on Consumption, Growth and Utilization Indices

It is well known from the different literatures that the mortality, growth, fecundity and other of the vital statics of insects vary widely with environmental factors. The relative small changes in the physical or biotic environment may have significant effects. The amount of food consumed by ***Bombyx*** larvae varies widely with temperature. Taking the amount consumed in a given time at 22 C as 100 per cent consumption fell 46 per cent at 14 C and rose to 203 per cent at 32-33 C. The A.D. did not vary greatly from 18 to 32 C but it was 44 per cent lower at 14 C.

Effect of Water Content of the Food

Legay (1957) found that ***Bombyx*** larvae fed mulberry leaves containing at least 70 per cent water retained much of the ingested water. About 30 per cent was passed out with faeces,10 per cent was lost as vapour and 60 per cent was stored.

Variation with Age and Sex

Age

Approximate digestibility and efficiency of conversion apparently do not remain constant, but rather vary with age during the growth period of an insect. The A.D. in ***Bombyx mori*** varies as 47,44,38, 38 and 37 in different instar like 1st, 2nd, 3rd, 4th and 5th respectively. A.D. also varies with in instars. The dry wt. A.D. for ***Bombyx*** varies as follows in the 5th instar : days 1 and 2 40 per cent, day 3 47 per cent day 4, 42 per cent day 5, 39 per cent and on days 6 and 7, 27 per cent.

E.C.I. is also decrease with age as follows in ***Bombyx***

1st–31, 2nd–27, 3rd–25, 4th–25 and 5th–23 per cent while the E.C.D. of Bombyx is 66, 61, 66,65 and 61 per cent from the 1st to the 5th instar respectively.

The decline of the E.C.I. with age is at least partly the result of the concomitant decline of the A.D. comparison of Hiratsuka's (1920) indices for the five instars of ***Bombyx*** shows that the E.C.I.'S and A.D.'S decline together and at about the same rate. The E.C.D.'S however do not show a tendency to decline with age. Thus, at least with ***Bombyx,*** the decline in the digestibility and not to a decline in the efficiency of conversion of digested food.

Sex

The sexes may differ in the efficiency with which they utilize food, but whether or not sexual differences are of general occurrence is not clear. The A.D.'s of both sexes varied with temperature but the percentage of difference between corresponding A.D.'s of female and male remained almost constant. In 5th instar the A.D., E.C.I. and E.C.D. of male and female ***Bombyx*** larvae are similar, although the females seem to convert a little more efficiently than the males.

Chapter 8
Silkworm Nutrition

Silkworm nutrition, broadly speaking, refers to the mechanism by which silkworm can obtain and utilize the substances to construct or build body and sustain the life activities, including the quality and quantity of diet, rearing environment, physiological situation and so on, while narrowly speaking, just refers to the quality and quantiy of food.

Silkworm feeds on a single food-mulberry leaf,from which it ingests various nutrients to maintain or support physiological activities. The nutrients include protein, carbohydrate, fats, vitamins, inorganic salt and water. Besides these air is also essential substance for silkworm. The requirement to the nutritive material of the silkworm responds to its ages. The relationship between the silkworm's growth, development and nutritive material has gradually come to light on the basis of advanced biochemical analysis and the improvement of artificial diet, which provides a scientific theory for silkworm rearing.

The composition of mulberry leaf

The fresh mulberry leaf usually contains

Moisture 70 per cent–80 per cent

Dry matter 20 per cent–30 per cent

Of 100gms of dry matter

Crude protein 24 per cent–36 per cent

Crude fat about 3 per cent

Crude fibre 9 per cent–10 per cent

Soluble nitrogen free matter 43 per cent–53 per cent

Among which

Soluble carbohydrate 12–20 per cent

Ashy substance 7 per cent–8 per cent

Vitamins B group and C (very little amount)

Both the contents of crude fibre and soluble carbohydrates increase with the leaf maturity.

The protein in mulberry leaves is a mixture of several kinds of proteins, polymerized by about 20 amino acids.

The non protein nitrogen content compound is mainly free amino acids or amides such as aspartic acid, glutamic acid, asparagin and glutamine etc.

There are many kinds of carbohydrates in mulberry leaves including cellulose and soluble carbohydrate like glucose, fructose, sucrose, maltose, starch, pectin calcide, pentose, arabinose etc.

Mulberry leaves contain little lipid. Among them there are mainly chlorophyll, lecithin, neutral fat, and sterol.

The ash of mulberry leaves include all kinds of inorganic matter such as calcium, phosphorus, magnesium and so on.

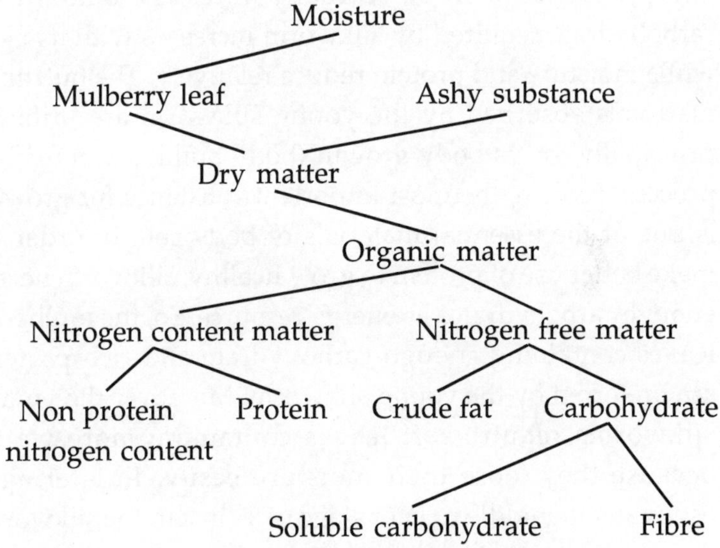

The mulberry leaf quality can be evaluated by experience and observation to distinguish its maturity such as softness lusture, thickness, smoothness etc. besides this by chemical method of analysis of the composition of leaves. It can also be identified by biological method with the rearing achievements to be the evaluation standards.

The Nutrients Required by Silkworm

Mulberry leaf is the only food for silkworm larvae. It is ingested and digested by silkworm for metabolism and to supply all kinds of essential nutrients for its growth and development. The minimum requirements of nutrients are basically meet from the nutrients present in the mulberry leaf. The nutrients content in the common leaf is higher than the requirements of the silkworm.

The tender leaf meant for young silkworms, especially for the first instar ones, is required containing more moisture and protein and proper carbohydrate. The amount of carbohydrate required by silkworm increases with its age, while moisture and protein reduce relatively. The nutritive materials absorbed by the young silkworm are utilized principally for its body growth (body build) ***i.e.***, nutritive process. Protein, the most important substance for growth is one of the essential materials of body cell. In order to make better use of protein to grow healthy, silkworm needs enough carbohydrates as energy resource. So, the mulberry leaves containing enough carbohydrate and rich protein are preferred by the young silkworm. Moreover the young silkworms require soft leaves containing more water because they loose their moisture easily. In later age, especially in middle or later stage of 5th instar, the silkworm needs more protein as materials to form silk substances, more carbohydrates to provide energy and to form glycogen and fat to store the nutrient. According to the physiological requirement stated above, it is advisable to use the leaf containing more water and protein, proper carbohydrate feeding for the young silkworm to use the mature leaves containing more carbohydrate and protein, proper water for the grown silkworm.

In order to maintain its growth, development and all life activities, the silkworm needs air to breathe as well as nutritive materials from mulberry leaves. Such as protein, carbohydrates, lipid, vitamins inorganic salts and water etc.

Air

Silkworm needs oxygen from the air to breath and to oxidise in its body to obtain energy for maintaining all kinds of life activities.

Carbohydrates

It is the essential energy resource for survival of the silkworm. It is one of the most important nutrients as proteins and fats. They often turn into glycogen to be stored in the body.

Carbohydrate content in mulberry leaf, varying with mulberry species, generally is 43 per cent–54 per cent relative to the dry matter weight, of which pectin and cellulose are above 10 per cent. It also contains sucrose, starch, glucose, fructose and pentose. Sucrose and glucose are most nutritive to silkworm, second is fructose. The digestivility of soluble glucose, fructose and sucrose is above 93 per cent.The utilization efficiency for starch is extremely low while dextrine and fibre cannot be utilized.

Carbohydrate Nutrition in Silkworm

Some sugars possess gustatory simulation effect on larval feeding of silkworm. Therefore, synthetic diet finds no use in evaluating the nutritive value of carbohydrates.

With the help of a hypodermic needle, compounds of carbohydrates either in solution or suspension were orally administered to a newly ecdysed unfed Vth instar larvae. Any carbohydrates which increased the survival beyond that of the control larvae given water alone were assumed to have nutritive value.

Another method used for evaluation of nutritive value of carbohydrates was to measure the rates of conversion of orally administered carbohydrate to blood trehalose and fat body glycogen since trehalose is the main carbohydrate of silkworm blood and glycogen is the storage carbohydrate (in fat body). Starved Vth instar larvae were orally administered with carbohydrate and 6 hours later the blood and fat body were collected for analysis.

In general pentose sugar were found to be of poor nutritive value. Among these xylose was found to be utilized slightly, where as arabinose, rhamnose and ribose was found to be of little or no value.

Hexose sugars like glucose, fructose was found to be utilized well followed by mannose. Glactose possessed a slight nutritive value where as sorbose did not have any effect.

Disaccharides, such as sucrose, maltose, lactose cellobiose, trehalose and melibiose were uniformly well utilized. Trisaccharides such as raffinose, melezitose were also well utilized.

Sugar alcohols such as sorbitol was of high nutritive value followed by mannitol. Inosital and dulcitol are of no value as source of carbohydrate.

It is considered that the requirement for carbohydrate might be satisfied by any one of the carbohydrates of high nutritive value. The utilization of carbohydrates in silkworm depends on the presence or absence of carbohydrases in the gut tissue in digestive juice. It is interesting to note that amylase is specifically active in the digestive juice and virtually no other carbohydrase activity is detected in the digestive juice. However, various glycosidases are found in the intestinal tissue.

Thus the digestion in silkworm proceeds in two steps: (1) Extracellular digestion and (2) Intracellular digestion.

Extracellular Digestion

Polysaccharides mainly digested in the gut lumen by the digestive enzyme present in digestive juice.

Intracellular Digestion

Disaccharides and trisaccharides are split in their constituent sugars in the gut tissue.

Mutant Strains

Completely lack digestive amylase. Oral administration of starch or dextrin were found to increase significantly the total blood sugar, blood trehalose and fat body glycogen in larvae having high amylase activity in the digestive juice but no significant increase was seen in case of mutants.

But when mutants were fed with mulberry leaves both blood sugar and fat body glycogen increased significantly. This proved that carbohydrates other than polysaccharides which occur in mulberry leaves are utilized well by the amylase free larvae, starch, dextrin–such polysaccharides are of minor importance in the nutrition of silkworm.

In the normal state of larvae the level of blood trehalose varies little during certain limited period of development and it is considered that there is then a homeostatic regulation mechanism. There are at least two distinctive components of trehalose in the muscle and midgut of the silkworm and their enzymatic properties have been clarified. When dietary glucose was increased markedly. Amount of blood trehalose corresponds to 20-35 per cent of the total storage carbohydrates of 5th instars larvae.

Carbohydrates and Feeding

Some sugars possess a gustatory stimulation effect on larval feeding of the silkworm. Sucrose stimulates feeding most followed by fructose and raffinose. Arabinose, galactose,maltose,melezitose and methylglucoside and inositol are more or less stimulating. Glucose is a very weak stimulant than sucrose at the same concentration.

However, the best result is obtained when mulberry leaf powder is served together with agar and sucrose.

Protein and Amino Acids

Protein is one of the essential materials of the body. It is the most important component of cell molecular structure, it is also the most important substance for survival. It can adjust and control the substance for survival. It can adjust and control the substance for metabolism and physiological function of the silkworms by combining the active substance like enzyme, hormones etc. with other substances. Different protein content in the mulberry leaf and different amino acids in proteins can directly affect the growth and development of the silkworm and the combination of silk substance. Analysis finds that the kinds of amino acids and their ratio in the protein of mulberry leaf and of cocoon silk are almost same. Above ½ of the protein and free amino acids absorbed from the leaves in the last instar are utilized to from the silk substance.

Silkworm larvae efficiently digest and utilize dietary proteins. An active protease is present in the digestive juice and its optimum pH is on the strong alkaline side. Recently the protease in the digestive fluid have been separated into three components and their enzymatic properties have been studied. The optimal level of dietary protein for maintaining good growth ranges from 20-25 per cent. The nutritive value of various kinds of proteins is largely dependent on their amino acids composition. Supplementation with limiting amino acids,markedly improves the nutritive value of gluten and zein.

According to nutrition and function of metabolism the amino acids (L–AA) needed for silkworm's growth can be classified into 5groups.

a) Essential amino acids (group 1): arginine, histidine leucine, isoleucine, lysine, methionine, phenyalanine, threonine, valine and tryptophane.
b) Essentiatial amino acids (group 2): Aspartic acid and glutamic acid.
c) Semi-essential amino acid: Proline.
d) Non–essential amino acids (group 1): alanine, glycine and serine.
e) Non–essential amino acids (group 2): tyrosine, cystine.

Essential Amino Acids

These amino acids are needed by silkworm for their growth, development and survival but they cannot synthesize or the amount of synthesis is very less and they must ingest from food, are called essential amino acids. These essential amino acids are same as those needed by other insects and advanced animals. Without any of them, silkworm cannot grow. Generally speaking, silkworm cannot synthesize these essential amino acids, but according to the experiment with isotope 14 C, it is possible for silkworm to synthesize some essential amino acids, although the amount is very less and cannot meet the need of silkworm.

The minimum requirement amount of 10 essential amino acids for silkworm (μg/g diet).

Amino Acid	*Content in Mulberry Leaf*	*Content in Silkworm Body*	*Minimum Required Amount*
Arg.	97	162	46
His.	40	84	32
Lie.	76	162	61
Isolie.	141	246	61
Lys	81	214	55
Met.	17	51	27
Phe.	76	112	48
Thr.	81	130	59
Tryp.	–	–	10
Val.	104	370	69

As shown in the table, except methione, the other essential amino acids in mulberry leaf can satisfy the silkworm's requirement. If one of the 10 essential amino acid is omitted from the diet, protein synthesis stops and the excretion of uric acid is enhanced. Phenylalanine and methionine can be partly replaced by tyrosine and cystine respectively. The conversion of phenylalanine to tyrosine and methionine to cystine via cystathionine had been demonstrated.

Proline is formed from arginine via ornithine and it can be replaced by either arginine or ornithine in the diet. On the other hand proline synthesis is inhibited by the addition of proline to the diet.

Phenylalanine and tryptophan seen to be synthesized from phenyl pyruvate and indole lactate although their formation is inadequate for the maintainence of good growth.

The omission of both glutamic and aspartic acid results in marked retardation of larval growth. The requirements of these acidic amino acids can partly be replaced by non-essential amino acids such as alanine, glycine and serine. Certain ratios of combinations of the amounts of acidic and non-essential amino acids are needed for optimal growth, and these ratios vary depending on the development of the insect and its sex.

In larval haemolymph, the amount of free acidic amino acids is very less, most of them exist in the form of amide.

In silkworm the following transamination is extremely active

Glutamine + Oxaloacetic acid = Ketoglutaric acid +Asparagin

Alanine + Ketoglutaric acid = Pyroracemic acid +Glutamine.

i.e., in silkworm body, aspartic acid and glutamic acid can be transformed mutually.

There are several reports of the composition of free amino acids in silkworm haemolymph. Large amounts of cystathionine and canthionine are found in the haemolymph. The quantities of these sulpher amino acids are much greater in females than in the males. This several difference seems to be associated with egg formation. Some silkworm mutants conspicuously accumulate a few amino acids such as leucine, valine in their haemolymph.

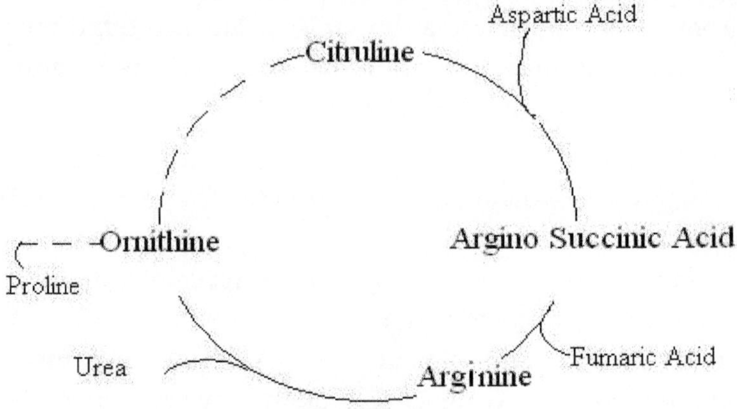

The Ornithine Cycle in the Silkworm.
The broken lines indicate absence of conversion.

Lipid Nutrition in Silkworm

The silkworm ***Bombyx mori*** is an oligophagus insect that feeds mainly on mulberry leaves. Much of the food eaten by the insect only during its larval life is converted into fats and lipids and stored in the cells of the fat body. Fats and lipids are especially appropriate as reserve energy source because of its favourable ratio of potential energy and metabolic water to weight. Insects characteristically accumulate it in high concentrations especially at physiological stages of development proceeding periods of non-feeding stage such as pupation and diapause or in maturing females for deposition in eggs. Apart from these major energetic functions fatty acids are the necessary components for phospholipids which are solely dependent on maintaining the structural integrity and functioning of probably all internal and external cellular membrane. Thus requirement of fatty acids falls into distinct groups

1) Structural and monosaturated fatty acids which occur as structural phospholipids but they are meant for energy production.
2) Polyunsaturated fatty acids which are essential nutrients and the insect are unable to synthesize them. The physiological reason for the essentiality of polyunsaturated fatty acids are unknown it may be significant that they tend to constitute a higher and stable proportions of fatty acids of phospholipids than of reserve triglycerides.

Requirements of Lipids in the Silkworm

Lipid nutrition in silkworm was first revealed by the experimental studies on the lipid composition of mulberry leaves, digestion experiment on mulberry leaves, lipid and feeding experiment with artificial diet.

Chemical analysis of mulberry leaves revealed that the lipid content ranges from 3 to 6 per cent on an average and the approximate digestibility of the worms were found to be 58.5 per cent.

Rearing on synthetic diet has proved that the worms require polyunsaturated fatty acids for their growth and development. No larval growth was observed if the ether extract were not added to a diet containing ether washed soyabean meal. Lack of these polyunsaturated fatty acids produced visible abnormalities in the adult wing. It was observed that the addition of mixtures of fatty acids to the diet was more effective than the addition of a single fatty acid in overcoming the deficiency. Subsequently the minimal optimal level of these fatty acids was determined.

In silkworm; myristate, palmitate, stearate oleate, linolate, linolenate, and arachidate were found to possess nutritive value. But linolenate and linoleate were found to be effective as deletion of both these acids resulted depression in growth. The dietary efficiency was greately improved with mixtures of these fatty acids as compared to single addition of either linoleic or linolenic acid. In mulberry leaf predominantly steoric,linoleic and linolenic acid occur in a ratio approximately 3:2:4 and together constitute 90 per cent or more of the total fatty acids.

Biochemical studies on silkworm showed that they are capable of synthesizing palmitic, stearic, oleic acid from glucose and acetate. The fatty acids were synthesized via pyruvate and decarboxylation. Thus the synthesis was found to be significantly affected by the dietary levels of fatty acids and carbohydrates. The omission of biotin from diet resulted in growth retardation and an alteration of fatty acid composition as a result of the inhibition of fatty acids synthesis

Dietary fatty acids from soyabean oil, corn oil, linseed oil were found nutritionally more effective in silkworm, compared to sunflower oil. Rape seed oil was found to be least effective in this aspect. The effectiveness of cotton seed oil, whale oil and cod liver oil was also reported. Futher more it was found that cholorogenic acid improves the nutritive effect of various oils and fats although the mechanism is unknown.

The fat bodies of Vth intar larvae and pupae were found to be rich in triglycerides where as the haemolymph contained neutral lipid consisting of diglyceridise. The lipid therefore may be transported by a mechanism similar to that found in ***Philosamia*** silkworm which the di-glycerides released from the fat body are transported by specific carrying proteins in the haemolymph. The presence of lipoprotins in the haemolymph of the silkworm discovered recently supports the above assumption.

Requirement of Sterols

The silkworm requires sterol as an essential nutrient for growth and survival and cannot synthesize the sterol denovo.

Experimental supplementation of cholesterol to mulberry leaves proved that the cholesterol has no nutritive value to the silkworm. Gas Chromatoghraphic analysis revealed that mulberry sterols consists mainly of □–sisterol with compesterol and stigma sterol as minor components which is similar to the sterol composition of most other plants.

The synthesis of cholesterol in silkworm was found to take place through dealkylation of □-sitosterol. This was demonstrated by using radioactive □(3H) sitosterol. The phytosterol are probably converted to cholesterol by the removal of the C_{24} alkyl group, which is a prerequisite to the utilization of these sterols in the silkworm. The conversion of □-sitesterol proceeds through the following pathway.

□-sitosterol → Fucosterol → Fucosterol-24, 28 epoxide

↓

Cholesterol ← Demosterol

The intermediates are found to be very effective in promoting growth of the silkworm. Stigmasterol and compesterol can also be used for growth.

The sterol may be required as membrane constituent and also as a precursor of moulting hormones as it is reported that Vth instar larvae can convert cholesterol to □ and □ ecdysone.

Furthermore, the sparing effect of fatty acids and vegetable oil for sterol requirements has been confirmed and the minimum optimal level of the sterol in the diet has also been determined.

Vitamins

Vitamins, a kind of organic active substance mainly are the component enzymes and other catalysts. They are essential substances to adjust the physiological function and metabolism in cells. The mulberry leaf contains B group vitamins which are necessary to the silkworm such as thiamine, choline, inositol etc. Among them, the requirements of choline and inositol are greater.

Vitamin C (Ascorbic acid) is also necessary to the growth and development of the silkworms. The diet value of the mulberry leaf increases with the content of vitamins which is greatly reduced in sunshine deficient leaves, over-long stored leaves, fermented leaves, which is also less in the autumn leaves than in the spring. If the absorption of vitamin is not enough, the growth of silkworm will be held back and dead silkworms even occur immediately.

Inorganic Salts

All cells contain inorganic salts which are of great importance to many life phenomenon activities of cells bioplasm and to the stability of cell bioplast colloid. The inorganic substances required by silkworm are K, P, Mg, Ca, Zn, Mn, Fe etc. The most important ones are K, P, and Mg.

Minimum Requirement of the Essential Vitamins (μ g/g dry diet)

Vitamin	*Minimum Requirements*
Choline	5.37
Inositol	5.55
B_5	0.08
B_6	0.02
B_2	0.015
B_1	0.001
C	113.55

The composition of inorganic salts in the body changes with the development, for example, Zn, Fe, Ca etc. declines dramatically after hatching. While K, Na,Ca, Mn increases with the start of feeding. More over in every tissue and organ, its composition is different. Regarding their physiological function–clear conclusion is not available. It is known that Zn is closely related to reproduction, the concentration of Ca and Fe in the body is linked to enzyme activity.

Water

In the silkworm the highest content is water which reaches 75 per cent–87 per cent. It is necessary to have that much water for the silkworm in all kinds of life process, especially in digestion. Water is the best solvent. Only in a solution state all kinds of nutrients can be absorbed into cells. Important chemical reactions in life activities take place only in solution. Moreover, only water can be used as the transportation of the nutrients and the waste matter. It can also adjust the body temperature and keep it relatively stable.

Chapter 9
Respiratory System of Silkworm

The respiratory system of silkworm is similar to that of other insects. It is a typical tracheal system; it supplies oxygen to the cells and removes carbon-dioxide from them. The respiratory system oxidizes the ingested food and produces energy to meet various activities of life such as growth, development and locomotion etc. if the respiratory system does not function normally, there would be detrimental influence on the normal metabolism and stable internal environment in the body. The physiological functions might be disordered; growth and development might be unhealthy.

Respiratory Organs

Respiratory organs are roughly divided into two spiracles and trachea.

Spiracles

The spiracles are apertures that open externally on the body surface, it is the passage for gaseous exchange, and meanwhile it can regulate water evaporation externally. In larval stage there are nine pairs of spiracles on the body, which are distributed on the lateral surface of the first thoracic segment and the segments from the first to the eighth abdominal segment, respectively. On the lateral junction (conjunctiva) between the meso-thorax and the

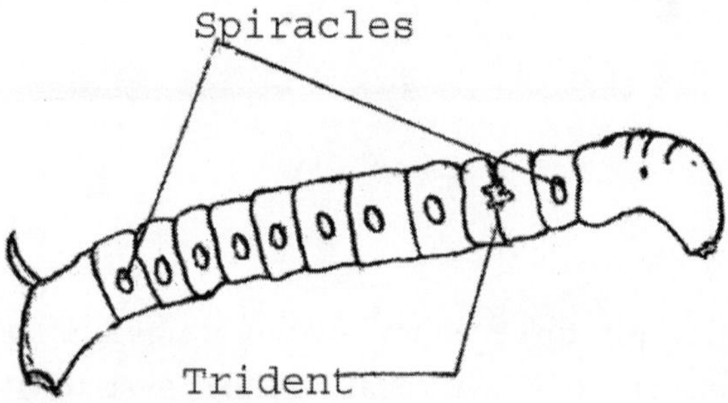

Position of Spiracles in Silkworm Larva

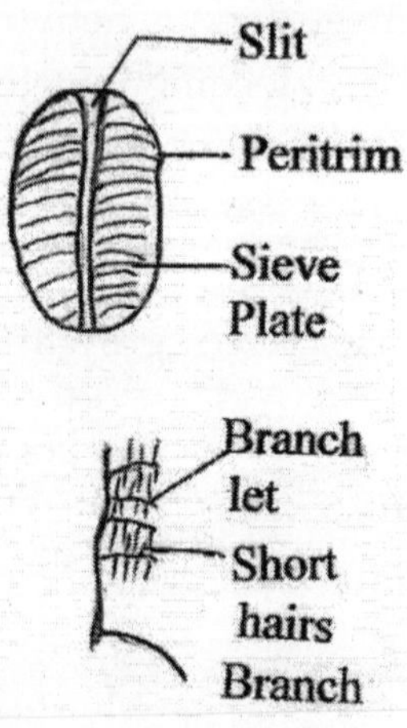

Detail Structure of Spiracles

meta-thorax, there is a trident aperture, which is a degenerated spiracle. It is function less for its poor development; however it has function during ecdysis.

The spiracles appear to be prolonged oval in shape which is enclosed by a black sclerotized peritreme; from the inner margin of the peritreme arises the sieve plate which meets at the middle of the spiracle to form a slit. The sieve plate is brownish and composed of many sclerotic branches. The numbers of branches increase with the growth of larva. Branches are grouped into 4-5 layers, and each branch bears many branch-lets from which arise short hairs. The sieve plate for the larvae in the fifth instar prevents the entry of foreign materials such as dust and pathogens, but allows the entry of clean air freely, and functions as a cleaner. The external opening of the spiracle leads into a cavity, called atrium, which contains the closing apparatus and tracheal bush.

The closing apparatus comprises of anterior membrane, posterior membrane, sub-posterior membrane, closing bow, closing bar, occlusor muscle and dilator muscle. Silkworm depends on this closing apparatus to regulate size of spiracular aperture and the time interval between opening and closing.

The tracheae are the passage through which air flows into the body. They appear to be brownish, circular, in cross section, and are flexible. A number of tracheal branches start radially from the spiracular cavity including the degenerated spiracle and form the tracheal bushes. There are also two large trachea called main or lateral longitudinal tracheal trunks, connecting all the tracheal bushes. And there is a ventral transversal trachea in each segment but the last connects the two tracheal bushes in same segment.

The trachea derived from the tracheal brushes ramify (divide) repeatedly throughout the body and organs, and ultimately each branch forms a star like cell called the end cell at its apex. The end cell ramifies into many finer tubules called tracheoles, which are less than 1 micron in diameter and carry oxygen directly to the cells and tissues. It is thought generally that the end cells are derived from the tubular wall cells. The tracheoles are enclosed by a thin layer of cytoplasmic sheath, and its tubular wall is very thin, some bear chitinous spiral thread, some do not. However, under electron microscopy, spiral thread in the intima can be observed clearly. The terminus of each tracheole is closed and oxygen supply is diffused from the tracheoles to the cells. Through such a network of trachae and tracheoles oxygen is carried to various parts, tissues and cells of the body and partial carbon-dioxide is removed through the same channels.

In adults there are eight pairs of spiracles on the lateral side of the prothoraic segment and the segments from first to seventh abdominal segments respectively. In addition, the degenerated spiracle in the larval stage keeps a residual marking at its original position. The closing apparatus in adult is similar to that in larval stage, the anterior and posterior membranes are absent except in the first pair of spiracles. The trachea are silvery-white in colour, with slightly iridescent, their taenidia are very thin. The tracheal distribution is similar with that of the larval stage, but there are nine tracheal bushes in adult and each tracheal bush ramifies into more branches than in the larvae; especially the tracheal system in the head is well developed, the parts of trachea between segments frequently swell and forms air sacs, so that ventilation could be increased greatly, the air entry into the tracheoles could be much easy and the air volume inspired and expired could be added significantly.

Micro-structure of Trachea

The outer covering of tracheal wall is the basement membrane, an acellular layer beneath which is the epithelial cellular layer, 1.5 to 1.7 microns in thickness, each cell appears to be multi angular,has a circular nucleus and pigmented granules in the plasma. The inner covering of the tracheal wall is the intima secreted by the cellular layer; intima is convex internally and thickens to form taenidia. In the lateral side melanin granules deposited in the intima could be observed. The taenidia are about 0.4 to 2.2 microns, and keeps the tube distended allowing air flow freely in the trachea.

The tracheal intima is renewed during each moulting. During the moulting period, the cells in the tracheal walls are active, chromatic bodies and chromatic threads could be observed in their nuclei, mitochondria, exasperates endoplasmic reticulam, golgi bodies and many secretory vacuoles could be detected in their plasma, it is at this period that the new intima of tracheal wall, is formed, and the old intima are torn down and expelled from the spiracle.

Mechanism of Respiration

The tracheal respiration depends mainly on the diffusion and ventilation. The respiratory system transports the gases through the trachea and depends upon the tracheoles to make gaseous exchange with the tissues.

Gaseous Diffusion

The gaseous diffusion takes place mainly because of difference in gaseous concentration between the atmosphere and tracheoles. With the constant consumption of oxygen in silkworm body, the oxygen concentration in trachea gets lower and lower than that in the atmosphere, thus oxygen in the atmosphere diffuses into the tracheal cavities. On the contrary with the constant increase of carbon-dioxide in the body, (due to metabolism), the carbon-dioxide in trachea gets higher than that in the atmosphere, thus carbon-dioxide diffuses externally. The trachea and tracheoles are distributed so wide and are so many in number that the diffusion meets life requirement even if there is difference of 2-3 millimeters of mercury between the tissues and tracheoles. The diffusive power of carbon-dioxide is much stronger than that of oxygen, about 1/4 to 1/3 of that of carbon-dioxide produced by metabolism, diffuse through the integument rather than the spiracles.

Gaseous diffusion is influenced not only by the difference in gaseous concentration between the inner and the outer, which is caused by the metabolism but also by the size and length of the trachea. In the grown larval stage, the trachea gets longer with growth of the larvae, gaseous exchange by diffusion gets more but difficult, furthermore, the ratio of the spiracle to body surface area gets lower, over growing silk glands presses the trachea greatly while the respiratory intensity gets higher, all these factors cause deficiency of oxygen supply. Therefore,it is important for the rearing house to be kept well ventilated so that the requirement of oxygen for silkworm could be met.

Tracheal Ventilation

The ventilation is caused by the difference in air pressure produced through the movements of muscle, the spiracular closing apparatus and the tracheal elasticity. The trachea themselves are expansible, when the trachea constrict; the tracheal volume may be reduced at 20 to 30 percents. When the ventral muscles relax, the spiracles open, the trachea expand because of their elasticity, the air pressure in the tubes is lower than outside the body, thus air enters the tubes, contrarily, when the spiracles close and the abdominal muscles constrict after the air entry, the tracheal volume is reduced greatly, thus air in the trachea is forced internally to enter the tracheoles. At this moment the spiracles reopen, but he body is kept under contraction, the carbon-dioxide in the body is expelled externally. However, the ventilation can only promote gaseous exchange in larger trachea, the gaseous exchange between the tissues and tracheoles depend still on diffusion. The diffusion is the major way of gaseous exchange for respiratory system in silkworm, while ventilation is nothing but an accessory respiration.

The ventilation is also influenced by the pulsation of dorsal vessel and the peristalsis of the alimentary canal. The strength of the ventilation is closely related to the development stage and active condition. A male adult on shaking wings has much stronger ventilation than a pupa.

Role of Terminal Tracheoles

The gaseous exchange between the tissues and tracheoles is regarded mainly by the osmotic pressure of tissues which enclose the tracheoles. Tubular wall of the terminal tracheoles is very thin, and permeable not only to the gases but also to the fluid, thus the terminal tracheoles are frequently filled up with fluid. The gas in the tracheoles can not enter directly into the tissues, and have to depend upon the osmosis of the fluid to diffuse. When the tisues

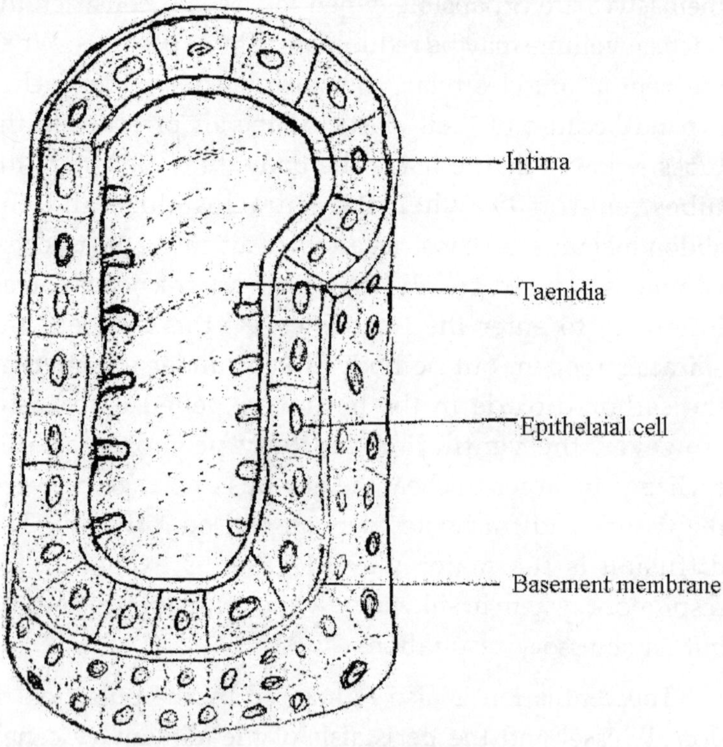

Histological Structure of Trachea of Silkworm (Diagrammatic View)

are active, for instance, when the muscles contract, the concentration of the intermediate metabolites in tissues is higher and the osmotic pressure in the tissue rises, as a result, the fluid in the tube permeates into the surrounding tissues, meanwhile, the gases in the tube reach the terminal tracheoles and oxygen diffuses towards the tissues. After the intermediate metabolites in the tissues have been oxidated, and osmotic pressure in the tissues has fallen, the fluid in the tissues re-enters the tubes because of capillary action, at the same time, carries parts of carbon-dioxide

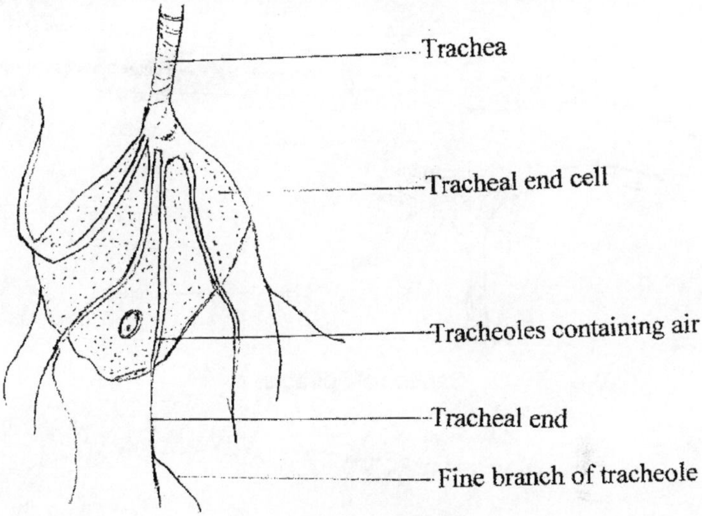

Tracheal End Cell

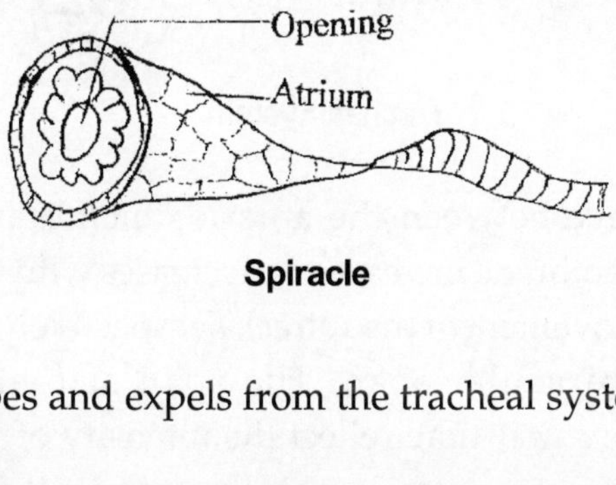

Spiracle

into the tubes and expels from the tracheal system through diffusion.

The ventilation is closely related with the water loss in the body. Respiratory activity is regulated by the opening or closure of spiracles, while ventilation is controlled by the change of respiratory strength and respiratory frequency. The combination of these two aspects makes the gaseous exchange for the silkworm body effectively, furthermore, keeps water balance for metabolism in the body.

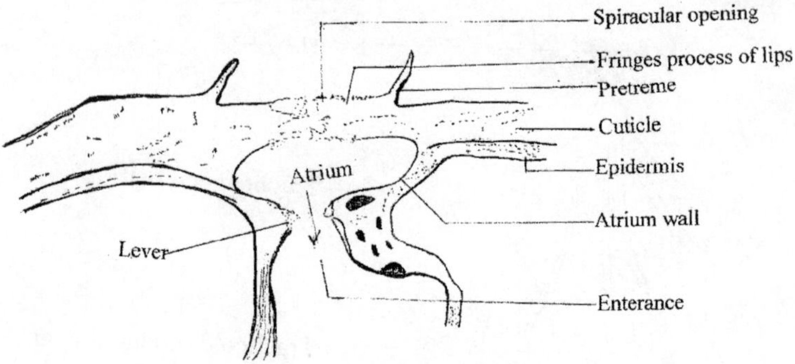

Section of Spiracle

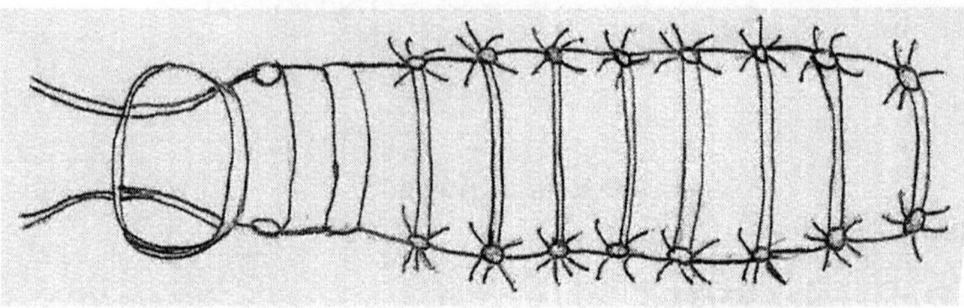

Tracheal System

The space between the anterior membrane and the posterior membrane increases or decreases with the opening or closing movement of the spiracles respectively. This space is called spiracular span. The span size and closing frequency per unit time reflect the intensity of respiration. And the span size varies greatly among the spiracles. Among the nine pairs of spiracles, the last two are largest, the span between the two membrane reaches to 140 to 210 microns, and the span of the first spiracle is the smallest, and reaches only to 30 to 98 microns, those of other six spiracles reach 70 to 182 microns. As to the opening counts, the first pair of spiracles is number one they open 19 to 27 times per minute under the temperature of 24 C, the last two pairs of spiracles open 3 to 20 times per minute, other spiracles seldom open,even some of them keep the spiracle closing all the time. The moving setae of spiracles in the in the 5th instar is different from the early stage and the gluttonous stage, spiracles open frequently at new stage 4 to 5 days late after the moulting, almost all of the spiracles keep open. This is mainly for the rapid growth of silk glands after middle stage of 5th instar and from the larger requirement of oxygen for the development of the silkworm.

Besides, deficiency of oxygen or surplus of carbon dioxide and accumulation of acidic metabolic products will lower the þH value and the electric conductivity of body fluid,stimulates the respiratory nerves to open the spiracles and to start ventilation. On the other hand, it is known that the carbon dioxide in the atmosphere might stimulates directly the closing of apparatus of spiracles, to open the spiracles by relaxation of closing muscles resulting from the lowering action potential from them. In the larval stage when the concentration of carbon-dioxide reaches to 4 percents, the spiracular span reaches to 100 percents, which cause the larva to increase the water evaporation. If the larva is put under the concentration of 12 to 15 percents of carbon-dioxide, it will expel gut fluid from its mouth and even it might be killed.

Chapter 10
Circulatory System

The circulatory system of the silkworm is an open one. When the blood moves in the body, the blood moves with fixed orientation only in the dorsal vessel and flows freely among the body organs and tissues, other cells and tissues can exchange for water, inorganic and organic substances with the fluid. Thus, the blood transports the nutrient, inorganic salts, water and hormones to various organs and tissues, mean while bring the metabolites to excretory organs and other tissues for intermediary metabolism and elimination.

The blood also serves as regulation of osmotic pressure, ion balance and pH value for various normal physiological metabolisms. Body fluid is the tissue which can transfer the pressure, and is helpful in ecdysis and emergence. The haemocytes provide the major defence mechanism against foreign materials both living and non-living which enter the body, and play important role in wound repairation, plasmagelation and immunization.

The dorsal vessel in larva is derived from anterior cell group of mesodermal coelom sac, it runs from the head to the ninth abdominal segment along the dorsal median line of the body, It is composed of two parts: the aorta and heart.

The aorta is the anterior portion of the dorsal vessel, and is a slender tube, its anterior end is aperture, which opens right behind the brain, its posterior end link with the heart at posterior portion of the first thoracic segment to the ninth abdominal segment where it is blind posteriorly, it is a pulsating organ associated with circulatory system, which causes the flow of the body fluid.

Each of the segments from the second thoracic segment to the ninth abdominal segment contains a pair of ostia on the lateral surface of the dorsal vessel, which is the passage of the body fluid entering the dorsal vessel and is incurrent. The margin of ostium forms into ostial valve by folding inward; therefore, it permits body fluid to enter the heart but prevents the fluid from flowing out.

There are eight pairs of alary muscles on the lateral side of the ventral heart wall from the second abdominal segment. One of their ends is attached to the vessel wall, and the other is connected with the integument above the spiracles. The alary muscles not only keep the dorsal vessel on its position, but also play accessory role in the heart pulsation. In addition to the alary muscles, there are lot of filaments which connect the dorsal vessel with other organs and/or tissues such as alimentary canal, tracheal trunks, fat body, integument and its muscles in or between which nerves, tracheal branches and microtrachae are distributed.

Histological Structure of Dorsal Vessel

The wall of dorsal vessel is composed of three layers: the outer membrane, middle muscular layer and the inner membrane. The outer membrane is a connective tissue, which usually contains few fibriform cells, thin web-like fibrillae, loosely distributed filament and fine granular membrane, the last encloses the tracheoles and sometime encloses the nerves. The muscular layer is the thickest layer, and is made of striated muscle fibrils which are ranged obliquely; there are nuclei unequal in size under the sarcolemma. Many glycogen granules are distributed in the plasma between the nuclei and muscular fibrils, lot of big mitochondria are distributed in the sarcoplasm. The muscular relexation results in the thinning of the tubular wall and expanding the tubular cavity. The inner membrane of the dorsal vessel is acellular membrane of the tubular cavity. In addition, there are pericardial nephrocytes on the surfaces both outer and inner wall of the dorsal vessel and around the alary muscles. The pericardial nephrocytes on the inner wall of the dorsal vessel loosely distributed, and cannot form a continuous layer.

Pupa

In the pupa the ostia have posterior valve flaps only and the backward flow of blood is not restricted, entering blood goes forward or backward.

The Dorsal Vessel in Adult Stage

In adult, the dorsal vessel extends from the head to the seventh abdominal segment along the dorsal median lines

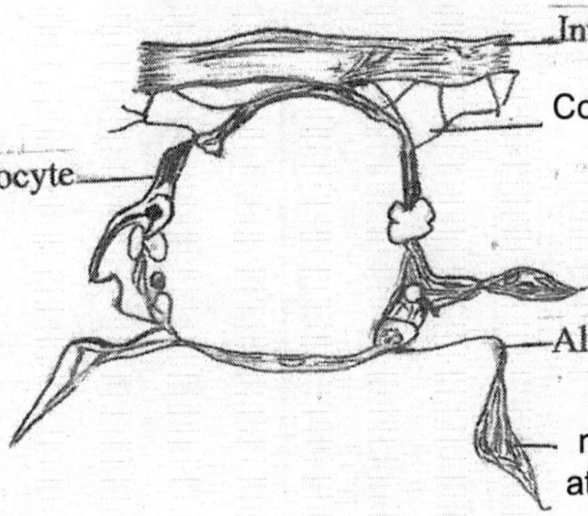

T.S. of Dorsal Vessel

of the body. It is closed posteriorly while the front end opens a little, before the brain. The portion located in the head and the thorax is slender tubular, called the aorta, while the posterior portion located in the abdomen is a more or less swollen tube, called the heart, which is adhered closely beneath the dorsal integument. After entering the hind thorax, the dorsal vessel curves sharply downward and extends upward the dorsal wall at the mesothorax, and then, curves downward and enters the head along the dorsal surface of the oesophagus, and opens a little before the brain. In adult there are only 10 pairs of ostia. However, there are still 8 pairs of alary muscles, which are enlarged and prolonged toward the laterals because of the swollen abdomen of the adult.

Chemical Composition of the Plasma

The body fluid is also called the haemolymph and it includes a cellular fraction–the colourless haemocytes and a liquid component–the plasma.

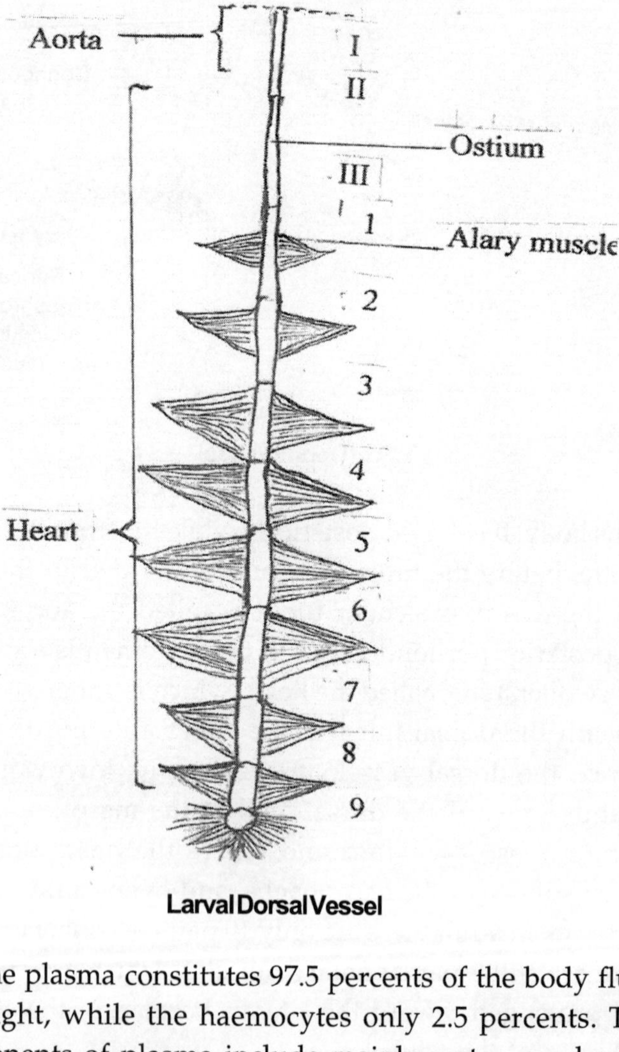

Larval Dorsal Vessel

The plasma constitutes 97.5 percents of the body fluid in weight, while the haemocytes only 2.5 percents. The components of plasma include mainly water, saccharide lipid, inorganic salts, organic acids, vitamins and large amount of carbon-dioxides and other metabolites as well.

Haemocytes

Blood cells are also called blood corpuscles, or haemocytes. Haemocytes count range from 1000 to 2000 haemocytes per milliliter. Haemocyte counts are less at new egg stage, but they increase with the growth of the silkworm, the counts are reduced greatly in the adult. At each instar, haemocyte counts are largest during moulting, after moulting haemocyte count reduced gradually until glutinous stage. Several types of haemocytes have been recognized–such as proleucocytes, plasmocytoids. Haemocytes play an important role in the defence mechanism against micro-organisms and in the encapsulation of the foreign bodies. Proteins, amino acids, salts and large amount of water are also present in the blood. The pH of the blood is approximately 6.3 to 6.5 and the specific gravity is 1.02 to 1.04. The main enzymes of blood are tyrosinase, amylase and catalase. Races of silkworms that spin white cocoons have pale yellow blood, where as races that spin coloured cocoons may have pale to deep yellow blood. Pigmentation of the blood is due to the xanthophylls in the mulberry leaf. The respiratory function of the blood is insignificant and mainly serves as a medium of transport for nutrients, hormones and metabolites.

Chapter 11
Excretory System

Excretion is the process by which an organism maintains its external environment within physiologic limits by eliminating the excess metabolites. This process operates in all cells, and the enhancement of this function in cells assembles into an excretory structure, or organ ***i.e.***, excretory system of the organism.

Excretory System of Bombyx

The Malpighian tubes are the primary organs of excretion in ***Bombyx mori***. It starts from the junction of small intestine and colon in the hind gut of the alimentary canal. The base of the malphigian tubes is enlarged like a baggy shaped — the urinary bladder.It branches soon into two tubes, one of which branches again into two therefore, finally three tubes are formed on both ventro-lateral sides of the intestine. These tubes stick to the mid gut and runs towards the anterior side. At around the 3rd or 4th abdominal segments they turn backwards, proceeding in zig-zag course around the colon and finally they enter the wall of the rectum. The malpighian tubes in the wall of the rectum proceed in winding, being separated into two lines by a thin membrane, inner lined tube and outer lined one. Finally they go blined. Distally they are usually free but in some insects the distal ends of the tubules are closely applied to the hind gut. This cryptonephridial condition occurs in the larvae and adults of many coleopteran, symphylan larvae, in myrmeleonted larvae and in almost al Lepidopteran larvae. These tubes are lie below the muscle layer of the

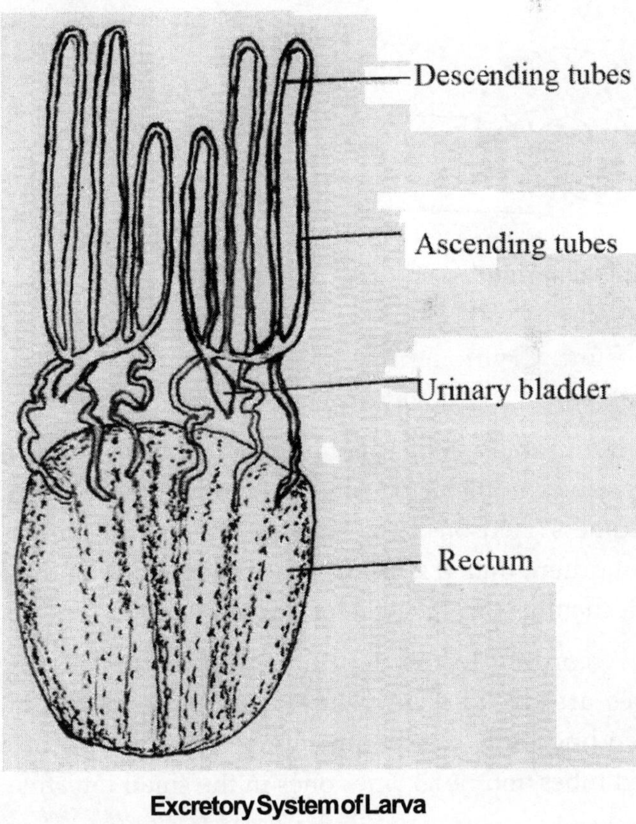

Excretory System of Larva

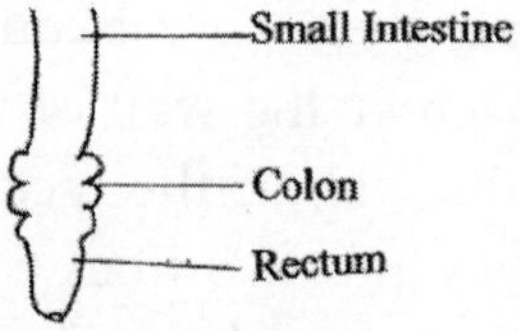

Posterior Portion of Alimentary Canal

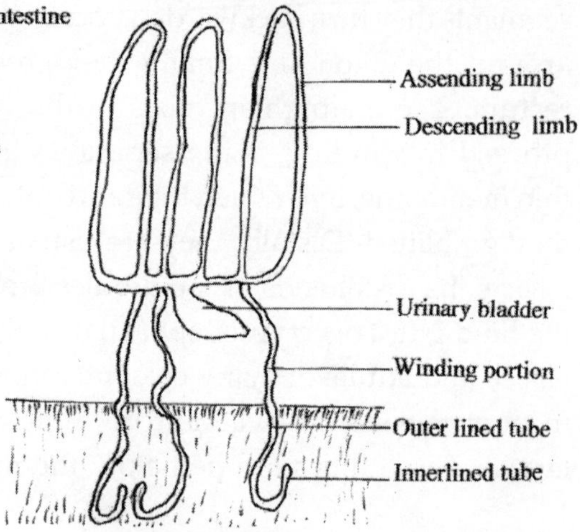

Malpighian Tube of Silkworm

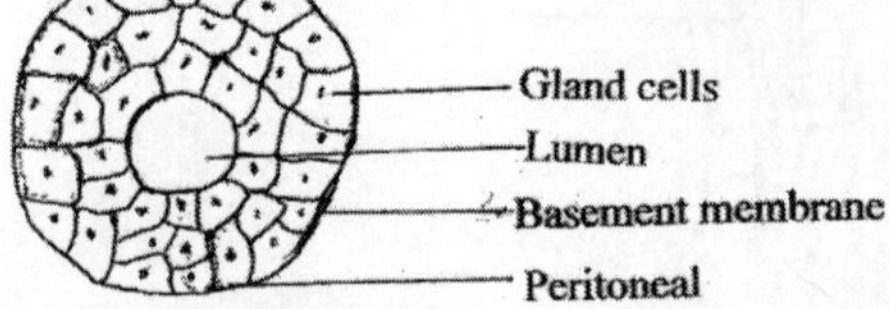

Section of Malpighian Tubules

hind gut. Cryptonephry probably enables the insect to conserve water by withdrawing its from the faces.

Malpighian tubes have generally been regarded as derivatives of the ectodermal proctodum by some authors but others (Henson- 1937-46) believe them to originate from an undifferentiated zone of cells between mid hind gut and treat them as out growths of the mesonteron.

According to the morphological feature Malpighian tubes are divided into several parts, urinary bladder, ascending limb, descending limb winding portion inner lined tubes and outer lines ones in the small intestine wall.

Malpighian vessels consist of gland cells and basement membrane. Regarding the peritoneal membrane, which is said to cover the external surface of the vessels, there are several opinions, but at present no agreement has been found.

The secretion of malpighian tubes consist mainly of the crystals of uric acid. It is said that the white powdery crystals on the body surface of the newly molted larva are the secretion of malpighian tubules and those crystals flow out of the crack between the old tunica intima and the new one at the time of larval moulting. The amount of uric acid in the malpighian tubules is variable according to the age of silkworm, showing the least amount in the moulting age and the largest amount in the middle age of silkworm.

Urine of Silkworm Larva

Urine of silkworm larvae is produced in malpighian vessels and excreted inside the rectum; accordingly faeces are mixed with urine. The latest brown colour feces of mature larvae is thought to be rich in urine. The composition of urine is as follows

1) Ash: 24.78 per cent
2) Organic matter: 75.22 per cent
3) Total nitrogen: 12.35 per cent
4) Nitrogen precipitated by cupric hydroxide: 6.57 per cent
5) Water soluble nitrogen: 10.35 per cent
6) Ammonium nitrogen: 0.54 per cent
7) Amino compound nitrogen: 0.88 per cent

The ratio of nitrogen contained in each nitrogenous compound is as follows:

1) Water soluble nitrogen: 83.80 per cent
2) Nitrogen in uric acid: 50.77 per cent
3) Nitrogen in organic base: 8.83 per cent
4) Nitrogen in mono amino compounds: 2.35 per cent
5) Nitrogen in di amino compounds: 4.78 per cent
6) Ammonical nitrogen: 4.37 per cent

Among the protein compounds main substances are as follows (In 1 kg. of the said brown coloured faeces):

1) Uric Acid: 10.128 g
2) Histidine: 12.84 g
3) Leucine: 1.87 g
4) Alanine: 0.87 g
5) Phenyl alanine: 0.59 g
6) Valine: 0.53 g
7) Proline, Aspartic acid, Lysine, Serine Less than: 0.34g

Urine of Moths

Urine of moths is a physiological combustion during pre pupal stages for ten and several days. It is preserved in the rectal sac and excreted after emergence of moths. Accordingly, urine of moth is kept uncontaminated with other indigestive matters. According to a survey of Hatano the total nitrogen occupies 24.2 per cent of the dry matter of the urine of moths and among the total nitrogen 18.5 per cent belong to the nitrogen of uric acid. From this fact it is calculated that urine of moths consists of 61.15 per cent of uric acid and 38.85 per cent of other substances.

Besides,ash components of the urine of moths are as follows (in order of amount contained in urine) phosphoric acid, sulphuric acid, potassium, magnesium, sodium and silicic acid.

Effects of Humidity on Urine

The amount of urine is directly affected by the relative humidity of the air. The water content of the feces is about 72 per cent in humid weather ***i.e.,*** on the rainy seasons where as it is only 35 per cent in dry air condition ***i.e.,*** in the winter season (Brussel, 1960)

Hormonal Control of Excretion

Altman (1956) suggest that the permeability of the malpighian tubules is affected directly by the hormones (types of hormones not identified). He suggested that the hormones secreted from corpus leuteum increases the rate of urine production where as hormone from corpus cardiacum causes reduction of urine formation.

Problems of Excretion

Mature larvae excrete soft and watery faeces during the period from mounting to starting of the spinning of cocoons. The amount of watery faeces is about 0.4 cc per capita, accordingly, in case 30,000 larvae (100 chandraki, 300 larvae per chandraki) are mounted in a room. The amount of watery faeces is about 12 litres. This water evaporates in the room, making the room moist. As the effect of the rising of humidity upon the quality of cocoons is comparatively less when the humidity is raised to 60 per cent to 70 per cent, however, the degree of lowering the quality of the cocoons is remarkable in case of raised humidity from 70 per cent–90 per cent. Sometimes, some cocoons are soiled or stained externally by the excreta of other larva. So, the excess humidity produced by the urination of mature larvae from their last excreta may be removed. The removal of excess humidity is done by the following methods:

a) Ventilation
b) Delittering in cocoon frames (in Japan it is known as Komonuki)

Ventilation

In case the outdoor air is moist like in a rainy day or night, the induction of outer air must be avoided and the elimination of moisture must be facilitated by using fire, opening the exhausting windows.

In case the outer air is drier than inside air, the elimination of moisture must be facilitated entering the dry air from outdoors, by opening doors, windows, inhaling and exhaling holes.

Delittering in Cocooning Frames or Komonuki

Soft faeces and watery feces (urine) excreted by mature larva are received on the mat just under the chandraki and they are taken away from the mounting room before water evaporates from the faeces. The operation is called delittering from chandraki or Komonuki. It is most effective means for eliminating moisture.

Delittering has to be performed just after larvae excrete faeces. It is effectless in case Komonuki is performed after the water has evaporated. Delittering is more effective in accordance with the rise of temperature of the mounting room.

Regarding the method of delittering, two straw mats or things like those are put just under the chandraki to receive the faeces. The upper mat with faeces, firstly is taken away 15 hours after mounting and another mat with feces secondarily is taken away 24 hours after mounting. By performing this method almost all faeces are taken away from the mounting room, eliminating the moisture of the room.

Recently the rotatory cocooning frame or the improved hanging cocoon frame are being used in great number in Japan, because, at the time of using these cocoon frames, the quality of cocoons can be improved by performing delittering timely and labour can be saved by adopting, the free mounting method.

The silkworm layings are solid or stained by the urine of female moth intended for laying egg, sometimes these nitrogenous waste products also helping development of micro-organisms on the egg surface. So for the elimination of the urine the female moths intended for egg laying are first placed on a piece of old newspaper which is taped to induce them to pass urine. After urination they are allowed to lay eggs.

Chapter 12
Reproductive System

The reproductive organs in silkworm include external reproductive organs and internal reproductive organs. The former originate from ectodermal layer embryonically, while the later originate from the mesodermal layer. In larval stage, the reproductive organs are in primary state, they mature gradually at pupation. The silkworm have their normal function such as mating and egg laying on the day when adult emerges out.

Female Larva

The larval internal reproductive organs of females are composed of three parts; 1) the ovaries, 2) reproductive ducts and 3) reproductive primordial. The ovaries are the structures where oogenesis occurs, and situated symmetrically in the fifth abdominal segment on two sides of the dorsal vessel. Each ovary is white semi transparent and triangular; from its outer lateral apex a slender

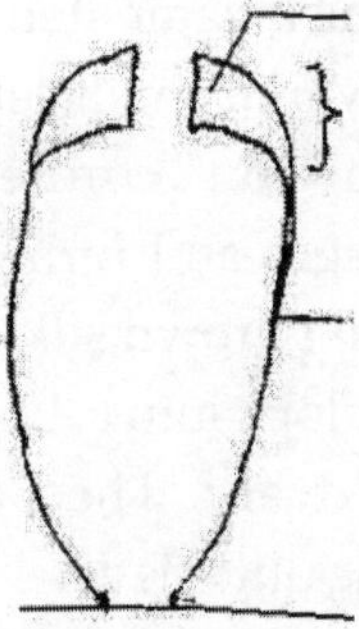

Female Larva

reproductive duct arises and prolongs backwards by the digestive tract and attaches to the integument of the VII abdominal segment. In addition, there are two pairs of reproductive primordia seen ventrally in the form of translucent spots on the eighth and ninth segments, respectively.

The pair on the 8th segment is called the Ishiwata's fore glands, while the pair on the 9th segment is called Ishwata's hind glands. The larval reproductive ducts shorten greatly during metamorphosis, transform into adult lateral oviduct and common oviduct. The Ishiwita's fore glands modify into the copulatory pouch, spermatheca and part of the ovipositor while the Ishwata's hind glands develops into ovipositor and accessory glands.

The outer covering of the ovary is the ovarial membrane. At the beginning of the newly hatched larva, the internal of the membrane has no obvious differentiation, two days later; the membrane invaginates to from the diaphragm.

While on the third or fourth day, the invaginated membrane has divided the cavity into four ovarial small cells (rooms). However, the membrane portion connecting the reproductive duct has no diaphragm and form a common cell called the basement cell. During grown silkworm stage each of the ovarial small cells develops into tubular ovariol, which grows irregularly in the ovary. The ovarial wall comprises of two cellular layers,the outer layer is the ovarial capsule while the inner layer is the follicle cells. The ovaries and ovarioles develop gradually in the larval stage. On the ovaries there are trachea which arises from the sixth tracheal bush.

Like the appearance of the ovary, the internal of the ovary also undergoes important changes during the larval stage, the oogonia in the ovarioles divide repeatedly with the development of the larvae. Apical cells (Wilson's cells) have appeared at the terminous of the ovarioles in the third instar. Oogonia beneath the apical cell divide energetically. Finally each oogonium forms one oocyte and seven nurse cells. These cells in group of eight move gradually down to the ovariole base, the differentiated nurse cells loose the ability to multiply, and serve to supply oocyte with nutrients. Among these eight cells, seven trophocytes gather above together to form a trophic chamber, while the oocyte forms a egg chamber. The trophic chamber and egg chamber are arranged alternatively in the entire ovariole,which, therefore, falls into the category of polytrophic ovarioles. The ovariol membrane is not broken until the second day after pupation. From that moment on, the ovarioles are free in the hemocoel, they develop further up to the late pupal stage, and the hemocoel is filled up with eight ovarioles.

Male Larva

The larval internal reproductive organs of males are composed of three parts: testes, reproductive ducts and reproductive primordia. The testes are the place, where the spermatogenesis occurs; their positions are similar to the ovaries in female, ***i.e.*** 5th abdominal segment, but they are larger than ovaries at same stage. They are milky white, semi transparent and slightly circular, kidney shaped, their opposite sides are concave, from where a slender reproductive duct arises respectively that prolongs backwards by the two lateral sides and connects to the reproductive primordial located on the posterior ventral centre of the 8th abdominal segment. The male reproductive primordia (Herold's gland) are larger at the anterior portion than at the posteriors portion, thus they appear to be pear shaped, their posterior ends attach to the anterior median integument of the ninth abdominal segment. The reproductive ducts are solid tubes except the portion near the testes, they become hollow tubes, develop into spermatic duct until pupal stage. The male reproductive primordia evolved into the seminal vesicle, ejaculatory vesicle, ejaculatory duct, accessory glands and aedeagus of the genitalia.

The outer covering of the testis is a layer of testis membrane, inside the membrane is the connective tissue,

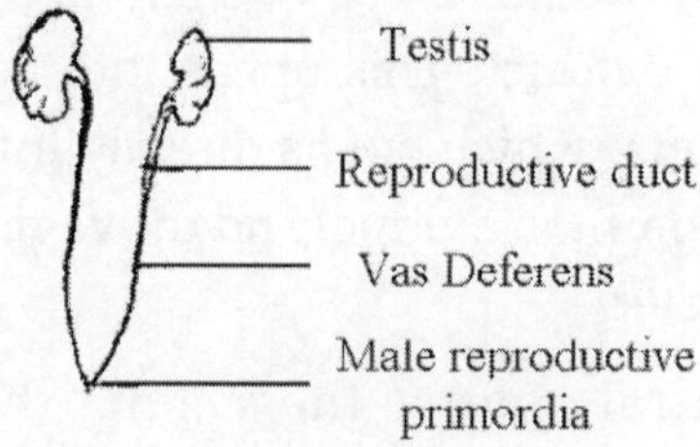

Male Larva

beneath which lies the intima composed of the flat cellular layer. On the second day after larva hatched from egg, the connective tissue forms the diaphragm, on the following day each testis is divided into four testicular follicles. Three invaginate sulci of the intima could be observed from outside aspect. Each testicular follicle has a testis basement cell (room), where no diaphragm exists, to form a common cavity which connects to the reproductive duct. However, the basement cell has no passage to the testicular follicle in larval stage.

Adult

Female

The female adult internal reproductive organs include ovaries, oviduct, vagina, copulatory pouch, spermatheca and accessory glands. Each of the paired ovaries comprises of four well developed ovarioles, each of which ends in a thin filament, four apical filaments of the same ovary bound together to form an ovarial ligament, which attaches to the integument. Slightly before adult emerges, the ovariole are filled up with the eggs in various stages of maturity.

The lateral oviducts are a pair of tubes, each of which connect to one ovary, and are the passage through which the eggs are discharged because of the primordial contraction of the oviducts. The two lateral oviducts join to form a common oviduct. This opens into an egg holding chamber, or vagina, which opens directly into the external ovipositor. The common oviduct and the vagina are uniform in tubular diameter.

On the ventral side of the vagina there is a oval copulatory pouch, its posterior portion is a thin duct, which

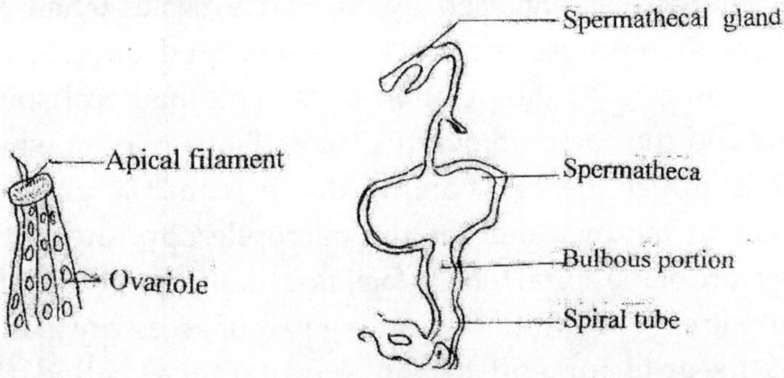

Enlarge View of Spermathecal Gland

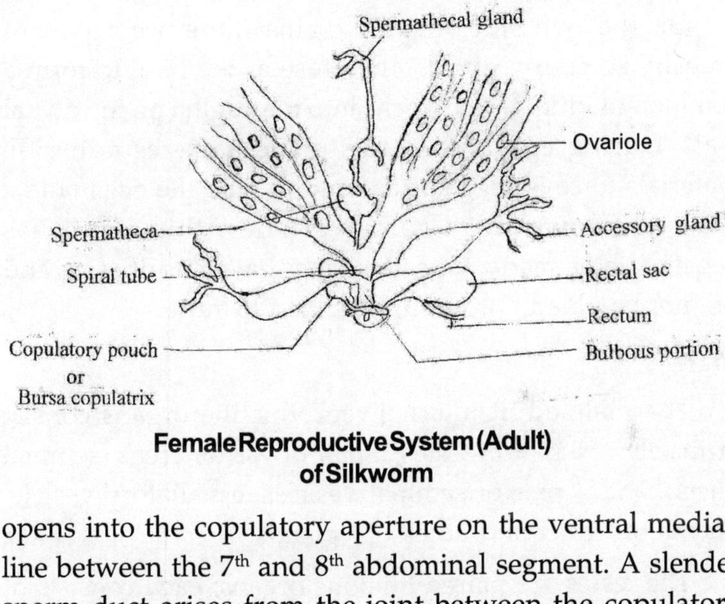

Female Reproductive System (Adult) of Silkworm

opens into the copulatory aperture on the ventral median line between the 7th and 8th abdominal segment. A slender sperm duct arises from the joint between the copulatory pouch and its duct. The sperm duct opens into the common oviduct.

A spermatheca arises from the posterior dorsal wall of the oviduct near the vagina. After copulation, the sperms are transferred into the spermatheca through the spermatic duct and common oviduct and are stored there. The spermatheca is enclosed by circular muscles,when the muscles contracts sperms are discharged outside the spermatheca. At the junction of the spermathecal spiral tube and the common oviduct is a bulbous portion which can support the egg coming down from the anterior common oviduct and has the micropile opposite to the aperture of the spiral tube to facilitate the entry of the sperm into the egg. In addition, there is a pair of accessory glands on the anterior end of the spermatheca, called the spermathecal glands whose secretion protects the sperm.

On the two sides of the vagina there are a pair of branchy accessory glands, they fuse at the base to form a common oviduct which opens into the vagina on the dorsal wall. During egg laying, this gland secretes adhesive materials to cover egg surface and to glue the eggs onto a support. However in some varieties, this gland has degenerated already, thus, their eggs have no adhesion and can not be glued onto a support or substratum.

Male

The male adult internal reproductive organs consist primarily of a pair of testes, a pair of vas deferens (seminal ducts), and a pair of seminal vesicles, ejaculatory vesicle, ejaculatory duct and accessory glands.

The testes are paired bulbous organs; each consists of four sperm tubes, or follicles, which are filled up with sperm bundles. The follicles have aperture opening into the basement cell.

The seminal duct is paired curved tubes, which connect to the seminal vesicles,is the passage that the sperms are transferred into the seminal vesicles,

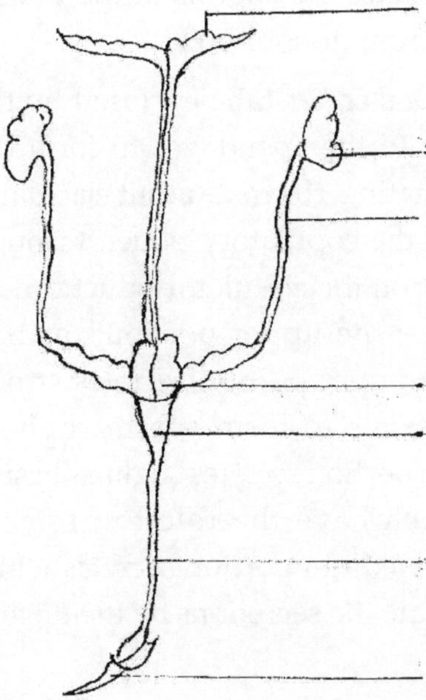

Male Reproductive System (Adult) of Silkworm

The seminal vesicles are paired capsular objects,in fact, they are posterior portions of the seminal ducts, but have swollen structure,their posterior ends lead to ejaculatory vesicle. The seminal vesicle temporarily store the sperms secrete some liquid materials to protect them and keep them resting in the vesicles.

The ejaculatory vesicle is also a swollen portion and is formed by fusion of two seminal vesicles. On its surface there are longitudinal sulci. The ejaculatory vesicle leads to the ejaculatory duct; its secretions have same function as those of the seminal vesicle. In the ejaculatory vesicle there are enough sperms stored for one ejaculation. The ejaculatory vesicle is enclosed by a strong muscular layer.

The muscular contraction makes the sperms in the vesicle ejaculate outside the body from gonophore.

The ejaculatory duct is a slender tube enclosed by the circular muscles, and leads to the apendages to form the male external genitalia. At mating, the male adult ejaculates the sperms from penis into the copulatory pouch through the ejaculatoy duct. In function the ejaculatory duct can be divided into three portions: the upper portion, median portion, lower portion. The upper portion secretes strong adhesive transparent materials to form spermatophore before ejaculation. The down portion secretes slight adhesive transparent materials which have the role to promote spermic activity. While the medium portion secretes white transparent materials to isolate the secretions by the up and down portion.

The anterior ends of the ejaculatory duct bear two functional accessory glands, most parts of which have fused their terminal ends blindly but keeps on isolation, their secretions from down portion are white liquid, which are mixed with the sperms to form the semen. The secretion not only supply the sperms with nutrients but also dilute sperms to facilitate their swimming. While the secretions from the upper portion are transparent and slightly adhesive liquid,which are secreted at the last copulatory moment to form a spermatic plug to lock the copulatory aperture of the female.

Chapter 13

Spermatogenesis in Silkworm

Spermatogenesis includes all successive processes, divisions and transformations undergone by germ cells in formation of spermatozoa.

In the embryonic stage, each testis consists of only one follicle in which a few primordial cells are observed. When the larva is about to hatch, the testis are partitioned into four testicular follicles by three membranes. At the blind end of each follicle, a giant cell is found, the so called apical cell, it is also known as Verson's cell. The apical cell is regarded as a supporting cell which connects all younger germ cells with the wall of the testicular follicle and probably nourishes them. The germ cells are arranged concentrically around the apical cell. In the newly hatched larva some tens of germ cells are present in the testis and their number gradually increases with the development of the larva. Kobayashi (1962) counted the number of primordial germ cells in the gonads of the youngest larva. The number of primordial germ cells increases suddenly after the time of hatching.

Number of Primordial Germ Cells during Hatching

Stage	*No. of Germ Cell per Gonad*
Two days before hatching	47
One day before hatching	59
The day of hatching	119
One day after hatching	171

Spermatogonia

These can be defined as the germ cells which arise from the primordial germ cells. Three days after hatching the primary and secondary spermatogonia are differentiated. The cytoplasm of the youngest spermatogonia (primary spermatogonia) is connected with that of the apical cell by a tubular projection. The more advanced spermatogonia (secondary spermatogonia) are contained in a common cyst termed as gonocyst. Intercellular bridges have been found by electron microscopic investigation. Each cyst contains a cluster of 64 primary spermatocytes and these are formed from a single spermatogonium after 6 cell divisions (Kawaguchi 1926).

Spermatocytes

Differentiation of the spermatogonia into the primary spermatocytes occurs first at the 2nd instar,differentiation of the primary spermatocytes from the spermatogonia occurs more frequently. It has been observed by electron microscopy that a canal system connects the sister spermatocytes. It was found that canal formation results from incomplete cleavage during mitosis which immediately preceeds meiosis. Pachytene spermatocytes appear in the late 3rd instar, and 10-11 days after hatching, the majority of cells in the testis are in the stage of synapsis and pachytene of the primary spermatocytes enter the 2nd contraction and diffuse stages, comparable to the diplotene stage in other organisms. The meiotic prophase persists until the larva reaches the 4th moulting or very early 5th stage 17 to 18 days after hatching,when cells in diakinesis and in metaphase-I are observed for the first time.Secondary spermatocytes and the first spermatids appear within a day after metaphase–I.

Synaptonemal Complex in Spermatocytes

Synaptonemal complexes were found in spermatocytes from young 4th instar larvae. Synaptonemal complex have been reported to occur in the meiotic prophase nuclei of the homogametic female, but not in the heterogametic male.However, the complex have been reported to occur in the oocytes.

Spermatids

Deriving from the 2nd maturation division, spermatids are haploid cells which, without further division and only through morphological transformation and modification of their components give rise to spermatozoa. This step is called spermatogenesis.

Spermatozoa

Appearance of spermatozoa begins on the day before the spinning stage of the larvae and the number of spermatozoa continuously increases until the 9th day after pupation. The number then gradually decreases since the mature spermatozoa pass into the vas-deferens.

Since the number of spermatozoa per bundle after 8 divisions of the spermatogonium is 256 and the number of spermatozoa bundles per testis is about 3000, the number of spermatozoa per head (two testes) is about 15.4 x 10^{-5}. The length of the spermatobundle ranges from 600 to 700 µ depending on the race of the silkworm. In the pupal testis, two types of spermatozoa are always distinguished. They are eupyrene and apyrene spermatozoa. In eupyrene spermatozoa, which are contained in a bundle in a cyst, a mass of thread like nuclei is seen in the anterior end of the bundle, while in apyrene spermatozoa cysts a number of small granule-like nuclei are scattered in the central part. Eupyrene spermatozoa develop from spermatocytes which have terminated the maturation division in the larval stage, while both types of spermatozoa are formed in the early pupal stage, and apyrene spermatozoa are formed in the late pupal stage as a result of poor nutrition. The apyrene

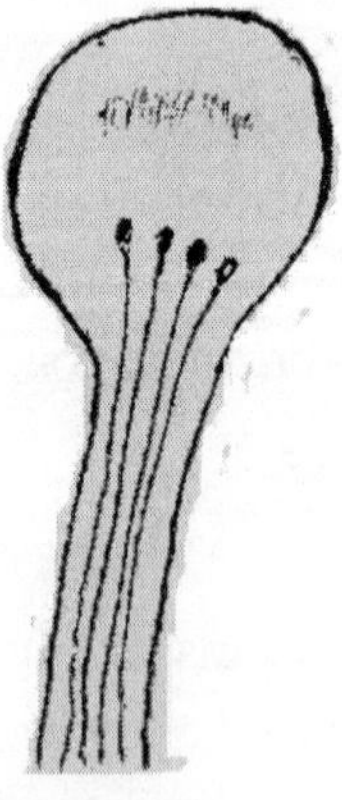

Eupyrene Spermatozoa

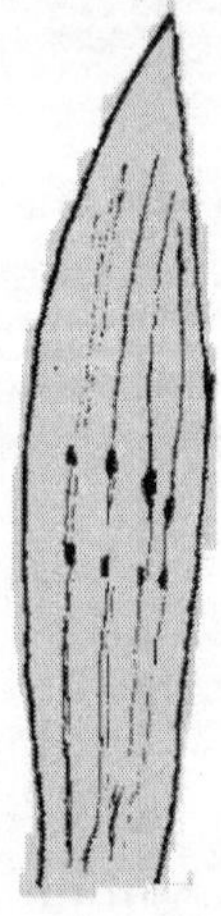

Apyrene Spermatozoa

spermatozoa have been believed by many authors to be useless in the process of fertilization.

From these results, together with those obtained from autoradiographic studies of spermatogenesis using tritiated thymidine, Sado (1965) estimated the approximate duration of each stage of spermatogenesis.

Approximate Duration of Each Stage of Spermatogenesis

Spermatogenic Stage	*Duration*
Primary spermatogonia	34 hrs
Secondary spermatogonia (sum of 6 generations)	5–6 days
Spermatocytes Resting Stage	1 day
Meiotic prophase I (Leptotene-Diakinesis)	9 days
Metaphase I to Anaphase II	1 day
Spermatid stage	9 days
Spermatozoa	1–14 days
Total Spermatogenesis	About 26 days

Dimorphism in the spermatozoa is distinguished in silkworms. The normal nucleated spermatozoa are known as eupyrene spermatozoa and the non-nucleated spermatozoa are referred as apyrene spermatozoa. The difference between the two is in the location and shape of nucleus. The apyrene spermatozoa are said to to be incapable of fertilization. Under natural conditions the apyrene spermatozoa are absorbed near the end of the testicular follicles. A nucleated sperm consists of a long filamentous body with an acrosome followed by the chromatic head, axial filament, seath substance and the tail. There is no nutritive cell near the head.

Chapter 14
Oogenesis

The egg or ovum serves two functions–it carries the maternal chromosomes and at the same time it provides the ground substance on which the future embryo develops. To be prepared for these double purposes, the egg during its development undergoes complicated modifications which results: (*i*) in production of haploid chromosome (nucleus); (*ii*) aciquisition of food reserves and (*iii*) preliminary organization of cytoplasm. The entire process is called oogenesis. Before discussing the details of the process it must be remembered that oogenesis begins at a stage when the individual itself is in embryonic condition and ends either shortly before or immediately after fertilization.

The process of oogenesis has been worked out in different organisms and the result shows close similarity in the process.

Stages of Oogenesis

The first stage of development of ovum is similar to that found in spermatogenesis. The egg mother cell or oogonia undergo proliferation by mitotic divisions. They then become oocytes and enter a period of growth. As the egg contributes the greater part of substances used in development, growth plays a much greater role in oogenesis than in spermatogenesis. Also the differentiation of egg occurs simultaneously with growth rather than after maturation divisions.

Nuclear Maturation or Changes in the Nucleus during Oogenesis

The nucleus enters into the prophase of the meiotic divisions. The homologous chromosomes pair together. But the subsequent stages of meiosis is postponed or suspended until the end of the period of growth of the cytoplasm. The nucleus of the oocytes increases in size but not as to the same extent as in cytoplasm. This increase in size is due to the production of large amounts of nuclear sap and the nuclei of advanced oocyte are filled up with fluid and often referred to as germinal vesicle. The chromosomes may increase in length but the amount of DNA in the chromosome does not increase in proportion to the enlargement of nucleus. Nucleolus also takes a greater role in oogenesis. Nucleolar material passes into the cytoplasm through the pores of the nuclear membrane.

After the oocyte completes its growth, reduction division proceeds. The nuclear membrane breaks up and the chromosomes are carried to the periphery of the oocyte. Here a spindle is formed perpendicular to the surface of the oocyte. Then the first maturation division occurs. Here the first divisions instead of giving rise to two secondary oocytes of equal size, one produces a very large cell- the secondary oocyte and one excessively minute cell called the polar body. In the second maturation division, the large secondary oocyte again divides into one large cell and one very small cell, the 2nd polar body. At the same time the small 1st polar body may or may not divide into two equal cells. In this way four ootids are produced (or three if first polar body fails to divide) of which the single large cell is the matured ovum and the smaller cells are polar bodies. These polar bodies are incapable of fertilization and degenerate. In some cases the 2nd meiotic division is postponed until fertilization occurs, ***e.g.*** The vertebrates as a whole. In Ascaris, Neries etc. both maturation division occur after the sperm has entered.

Cytoplasmic Maturation during Oogenesis

(*a*) Formation and Accumulation of Deutoplasm

The cytoplasm of the young oocyte is small in quantity with a clear homogeneous texture. As the oocytes develop the cytoplasmic and nuclear volumes increase and homogenicity of the cytoplasm is soon lost by the appearance of deutoplasmic substances. First of all the lipid droplets begin to appear, after that glycogen appears and finally yolk protein arises.

It is believed that the fat droplets and yolk spherules originate by the activities of mitochondria, golgi substances and of certain vacuoles, which are derived from special bodies called the yolk bodies. In general, two types of yolk bodies have been described: (1) yolk nucleus of Balbiani and (2) mitochondrial yolk body of Brambell.

(1) *Yolk Nucleus of Balbiani*

The yolk nucleus consists of a central body within which centrioles are found. This central body is surrounded by a layer of golgi substances and mitochondria (chondriosomes). In the early stages of the oocyte this yolk nucleus lies next to the nucleus but in later stages it breaks up and its fragments are distributed to the periphery of the oocyte, yolk platelates appear first and for this reason it has been concluded that the production of yolk is initiated by the components of the "yolk nucleus".

(2) *Mitochondrial Yolk Body of Brambell*

According to Brambell the yolk body is composed entirely of mitochondria, the golgi substances and centrospheres are absent. Yolk formation proceeds as follows:–As the young oocyte grows, the mitochondria increase in number and form the mitochondrial cloud. The transiting mitochondrial yolk body breaks up and dispersed into the cytoplasm. Some of these dispersed mitochondria transform directly into yolk spheres or nucleus.

(*b*) Golgi Bodies

In the younger oocyte golgi bodies are found around the centrosomes. In the mature oocyte golgi bodies sometimes disappear completely, possibly these organoids have been used up or transformed into some other structures.

(*c*) Mitochondria

Mitochondria show a tendency of aggregation in the yolk body but later they disperse in the cytoplasm. But unlike golgi bodies they never disappear from the growing oocytes.

(*d*) Endoplasmic Reticulum

It is rarely found in the growing oocyte.

(*e*) Morphogenetic Organization within the Cytoplasm of Egg

Two general categories of substances are developed within the cytoplasm of developing oocytes namely:

1) The visible or formed cytoplasmic inclusions and
2) An invisible morphogenetic ground substance.

The former group comprises the yolk spherules, fat droplets, pigment body etc. which can be seen with the naked eye or by means of microscope.

The later group ***i.e.***, the invisible morphogenetic ground substance probably is composed of enzymes, hormones and various nucleo-cytoplasmic derivatives enmeshed within the living cytoplasm.

The exact mechanism involved in the morphogenetic organization of the egg is not known. But embryological studies forced us to conclude that during the growth of oocytes, basic conditions are formed from which the future individual is formed or developed.

Chapter 15
Embryogenesis in Silkworm

In the eggs of silkworm as in most insects there is a distinction between the anterior and postertor poles which bears a definite relation to the position of future embryo. The eggs are located in the overioles in such a position that the cephalic pole of each is directed towards the head of the parent; also the dorsal and ventral aspects of the egg corresponds with those of the parent and of future embryo (Hallez 1886).

Fertilization of Silkworm Eggs

When male and female silk moths copulate, ejaculation occurs and innumerable spermatozoa enter the bursa copulatrix in the female reproductive system in the form of spermatophores in the bursa and ultimately reach the spermatheca or seminal receptacle. When the eggs pass down the oviduct prior to laying, the sperm moves from the spermatheca to the oviduct and enters the egg through the micropole, which is situated at the anterior of the egg. More than one sperm may enter the egg, but only one of them can fertilize the ovum, rest degenerate and disappear.

When the sperm enters the ovum and the eggs are laid, the first maturation division (reduction division) of the egg nucleus progresses from the middle stage to the late stage and is completed in about 40 minutes after egg laying. One part of the divided nucleus proceeds inwards as the egg nucleus and the other part becomes the first polar body and remain stationary. This is followed by the second part of the maturation division in which the egg nucleus divides, where as the chromosome number is reduced to "n" and the division results in formation of egg pronucleus and the second polar body. Meanwhile, the first polar body has also divided into two units.

The sperm which has entered the ovum becomes differentiated and it is only the head Part that moves towards the ovum. By the time the egg nucleus completes the second part of maturation division, the sperm pronucleus is formed which approaches the egg pronucleus. Subsequently the two unite and fuse to form the zygote. The chromosome number becomes double or is restored to 2n. Nuclear fusiion occurs in about 120-140 minutes after egg layig,when the eggs are preserved at 25 C.

The fused nucleus is near the anterior end of the ovum and it undergoes rapid and repeated divisions giving rise to a large number of nuclei. Some of the daughter nuclei move towards the peripheral parts of the ovum enter the periplasm and undergo further division. The other cleavage nuclei remain at the central part of the yolk and form yolk

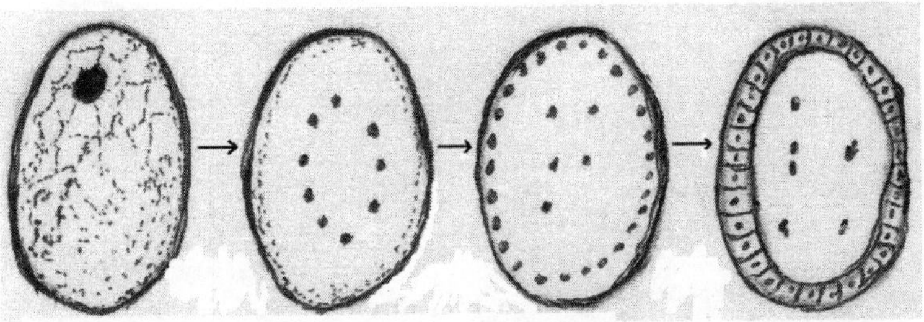

Cleavage

cells or vitellophages. Their function is to supply nourishment to the developing blastoderm cells.

A little later each of the nuclei in the periplasm is seperated by surrounding membrane to form a cell. Collectively these cells give rise to a cellular layer called blastoderm. This happens 15 hours after the egg laying. In this way, in case of silkworm eggs, at first nuclear division progresses followed by the formation of cellular layer around the periphery and then by the cleavage of yolk. This is because silkworm eggs are centro lecithal and this type of clevage is known as superficial clevage.

Formation of Germ Band

After the formation of the blastoderm, the cells on the the ventral surface of the egg become enlarged. This enlarged part is to form the future embryo and it gradually invaginates inside the egg. At the same time, wrinkles appear around the periphery of the enlarged part. The folded part gradually covers the enlarged part and becomes the amniotic fold. It is composed of an outer and an inner layer. The outer layer becomes connected with the residual

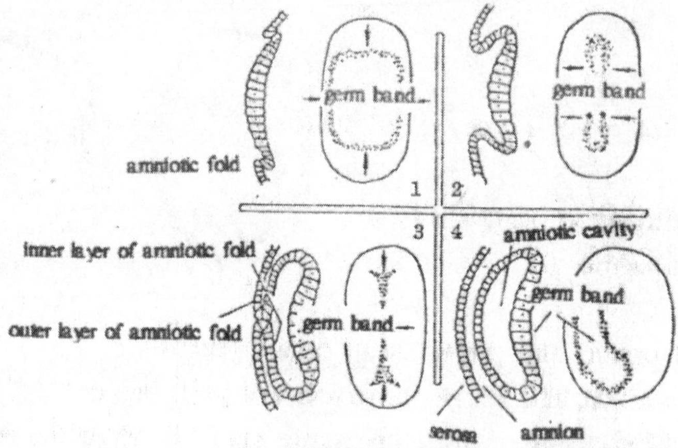

Germ Band Formation: Vertical Section

part of the blastoderm and forms a covering of the egg giving rise to chorion. The inner layer covers the outer surface of thickened part of the blastoderm and becomes the amnion. To attain this condition it takes about 20 hours after egg laying.

The germ band which has become differentiated appears in the form of a thin layer of cells in almost rectangular shape. Gradually it becomes elongated, followed by a right and left constriction near the anterior end. In this condition the embryo resembles the shape of "Daruma" (a traditional Japanese doll). The anterior enlarged part becomes the headfold or cephalic lobe and enlarged part becomes the tail fold or caudal lobe. This stage is attained about 24 hours after egg laying.

Differentiation of Germinal Layers

Along the median line of the germ band on the upper side a long narrow depression the primitive groove or streak or median plate is formed. The primitive groove is narrow at the centre but becomes wider at the two edges. Below the primitive groove a new group of spherical cells develop which ultimately give rise to the mesoderm, which can be distinguished from the columnar cells of the original tissue or ectoderm. Some of the median plate cells of ectoderm of lower layer penetrate deep into the embryo and differentiates to form the ectoderm cells later on.

The stages of development described slightly vary among different varieties. The following figures give the standard process of embryonic development when the eggs are preserved at 25 C after egg laying.

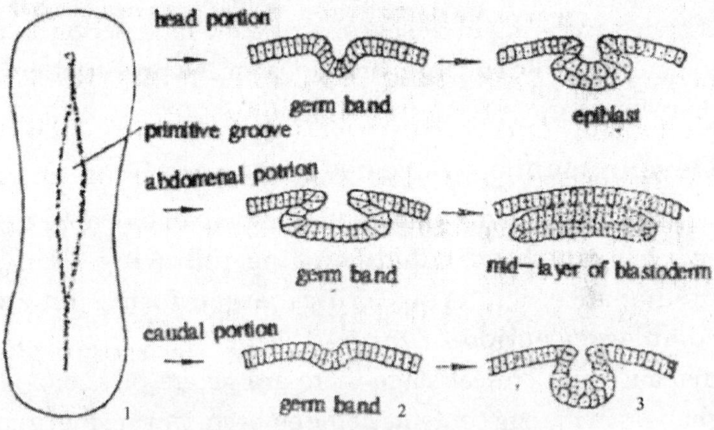

Ectoderm and Mesoderm Formation
1: Ventral view of primitive groove; 2: Early invaginated condition in primitive groove; 3: Mesoderm formation

The Growth of Embryo

Once the embryo attains the Drauma or guitar shape stage in ehe egg, the course of further development of embryo changes depending upon whether it is a case of diapause eggs or non-diapause eggs. The development of embryo of diapause eggs is delayed gradually and finally, becomes arrested. This condition is said to be the hibernating condition and once the embryo enters hibernation, unless it is allowed to pass the winter or kept at a suitable low temperature for some time, the embryo will not resume its growth even when suitable enviromental conditons are provided. On the other hand, the embryos of non-diapause eggs continue to grow at 25 C and hatch in about 10 days. There is much significant differences in relation to developmental biology and it is also impportant as far as silkworm egg production is considered. Since the eggs (during the period of hibernation) do not stop the physiological activities, it is necessary to know all the changes occurring in the eggs during the hibernation so as to determine the suitable conditions for the preservation of diapause eggs.

Development of Diapause Eggs

The course of development of embryo of diapause eggs can broadly be divided into the following period: prediapause, wintering, critical,organ formation and completion periods. During the period between pre diapause and critical stage there are severe physiological changes in the egg contents in the embryo. During the organ formation stage and completion stage there is much morphological change.

Pre Diapause Period, Diapause Period, Wintering Period and Critical Period

On the 3rd day after egg laying, the embryos cease further development. This stage is known as guitar and the

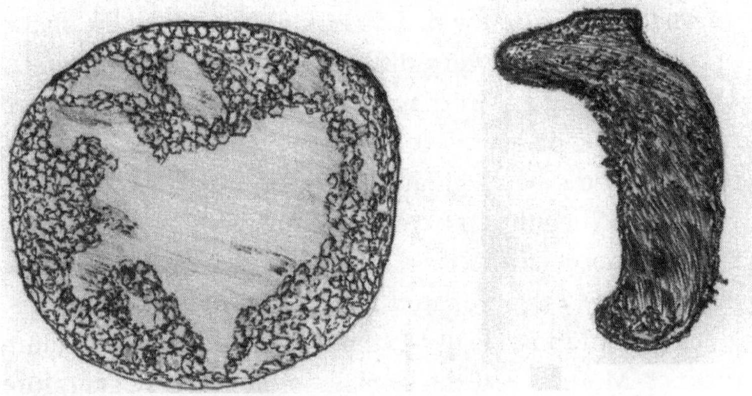

Embryo during Hibernation Stage

embryos attain the condition suited for hibernation. Since this period between the fertilization and guitar stage is before the entrance into hibernation, it is known as pre diapause period. After completing pre diapause period the embryo enters hibernation or diapause period.

The diapause period starts by about two weeks after egg laying. This period can be divided into period I and period II. During the diapause period I, the yolk cells present in the centre start moving towards the periphery region. Subsequently, the central parts become devoid of yolk cells and instead, a fluid filled cavity forms. The embryo is in the form of an accumulated mass surrounded by yolk cells. In diapause II stage the fluid filled cavity further enlarges and the yolk cells almost disappear. Another nucleoli of the cells of the embryo. The condition is developed in about 2 months after the egg laying.

The wintering period follows the diapause stage. This can be divided into wintering periods I,II,III and IV. In this period, the eggs which are stored at low temperature gradually break up the diapause. On the basis of the shape of the embryo, it is very difficult to differentiate between the period I and II in diapause stage. However,there are some differences in the condition of yolk, that is, most of the yolk forms an irregular mass or by the accumulation of granules. When the embryo enters the wintering period III, the mesoderm of the thoracic segments of the embryo is comparatively larger than that of the other body segments. During wintering stage IV, the mesoderm becomes more distinct. Moreover, if the eggs are stored at a temperature suitable for the development of embryo, the larvae hatch during this period. Hence,in order to obtain uniform hatching at a particular time, the development of the embryo has to be carefully monitored and controlled by preserving eggs at a lower temperature.

After the completion of the wintering period, the embryo enters a stage which is known as critical stage.

The development of the embryo at this stage becomes rapid once the diapause is broken and the embryos are kept at a suitable temperature. This period can be divided into I and II. During the critical period I, the head fold enlarges and the embryo becomes longer in shape. On the other hand the mesoderm becomes spread out on the left and right sides. During the critical stage II, the head fold further enlarges and the thoraic part becomes differentiated like a leaf bud from the head fold. The changes in the external appearance of embryo from the diapause to the critical stage are delicate, as a result of which, it is difficult to differentiate the various stages in between. However, as far as preservation of eggs is concerned, this stage is one of the

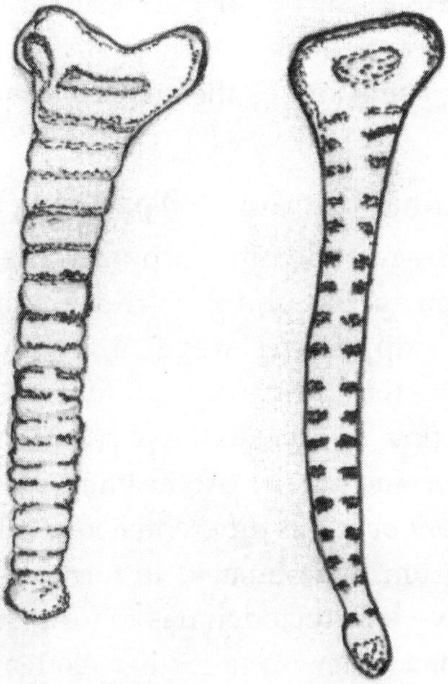

Embryo during the Critical Stage

important stages. During this period, the nucleoli of the cells of the blastoderm are larger than those during the diapause stage. This is also one of the characteristic features of this period.

Organ-Forming Stage and Completion Stage

Once the critical period is over, the embryo is conpletely out of the diapause stage. If preserved under suitable temperature, development is resumed and various organs start appearing. Hence, this stage is called the organ-forming stage and on the basis of morphological features, this can be divided in to 10 stages.

1) Neural Groove Appearing Stage

A neural groove appears along the midline of the embryo. The embryo gradually elongates and occupies about the three-fourths of the peripheral region of the egg.Thoracic appendages (Protuberances) appear in the thoracic segments and as the embryo grows they become longer.

2) Abdominal Outgrowth Appearing Stage

In this stage abdominal protuberances (appendages) appear even in the abdominal segment (Abdominal appendages appearing stage). Later, a pair of labial protuberences (outgrowth) appear in front of the head fold at the same time, stomodeum and proctodeum appear and gradually become tubular. At the blind ends of these ducts, the inner layer becomes differentiated and the new layer at the stomodeum is designated to form the future foregut while the layer of proctodeum is to form the hind gut. The newly formed inner layer further elongates to form the midgut.

3) Constriction Stage

At this stage the embryo gradually shortens. Of the 18 somites, 17 have a pair of spiracular sulcus each. The spiracular sulci of the somites no. 7, 16 and 17 disappear. The sulci of the somite no. 6 become vestigeal spiracles. The remaining forms the functional spiracles. Later the embryo further shortens.

4) Pre Involution Stage

In this stage, the first 4 segments fuse together to form head, while segments no. 5-7 forms the thorax. The rest of the segments become the abdomen. Of the protuberances, on the abdomen 4 pairs (no(s) 10-13) form the abdominal appendages and 1 pair (segment no. 18) become the caudal appendages. All the others gradually disappear.

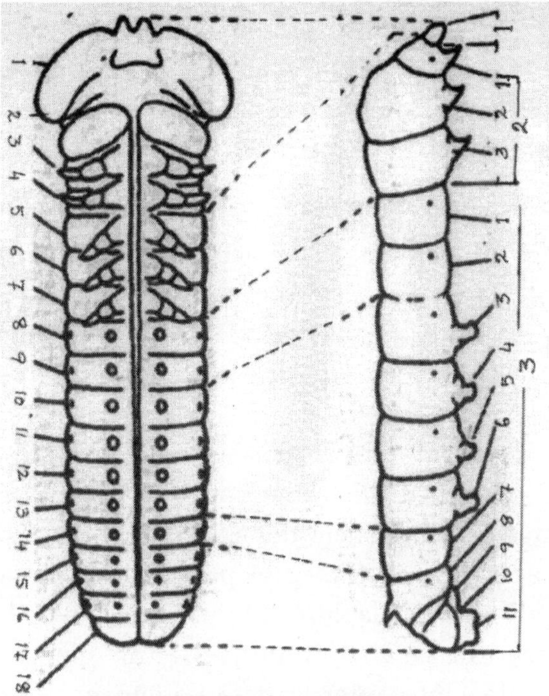

Segments of Embryo and Larva
1: Head; 2: Thorax; 3: Abdomen

5) Involution Stage

In the reversal stage the embryo starts to move around as it seeks its correct position so that its abdomen faces the inner side, this is finished in a short period which can be divided into 3— early, middle and later stages because of significant changes. The embryo increases in length once again as soon as the reversal stage is over. Hence, in order to ascertain whether the growth of the embryos is uniform or otherwise, this stage is treated as a landmark. Again with this stage as the borderline, the reactions of the embryo to the preserving temperature changes and hence it is very important in the context of blue colour (eye-spot) in the egg. The midgut does not yet form during the involution stage. However, the malpighian tubules have already appeared at the end of the proctodeum.

6) End Reversal Stage or Involution Completed Stage

After the involution is completed the formation of dorsal and ventral surfaces of the midgut progresses and the amnion gradually covers the dorsal surface coming from the right and left sides. The organs that form from ectoderm progress rapidly and the nerve cells become independent from the ectoderm and form nerve ganglia. The dorsal blood vessel also forms at the same time. During the accessory appendage forming stage the silk gland also appears as minute apertures in the basal part of the protuberance which will form the labium. At this stage the opening for silk secretion is also found and the gland can be differentiated into anterior, middle and posterior parts within 24 hours after involution. The reproductive cells form from the mesoderm and become accumulated giving rise to a rudimentary genital structure. Mesoderm also gives rise to muscles and fat tissues. In this way when the embryo attains the stage where important organs form, an opening on the dorsal side (nicknamed, the naval) becomes narrower and narrower and finally becomes a minute aperture on the dorsal surface at the junction between the segments no.2 and no.3 of the thorax. Finally the opening closes.

7) Tubercles and Bristles Appearing Stage

The embryo at this stage develops tubercles on the body surface. Bristles formed from the tubercles cover the body and the bristles on the dorsal side are found facing anteriorly and posteriorly with margin between the segments no. 2 and no. 3 of the thorax.

8) Trachea Taenidia Appearance and Colouration Stage

As the embryo grows further in this stage, taenidia appears in the trachea. The embryo also becomes coloured. Simple eyes form on the head and the tips of the mandible become coloured. The colour becomes gradually darker. Since the embryo moves actively during this stage, the chorion ruptures. As the embryo swallows the ruptured chorion and yolk, the digestive tract of the embryo also becomes dark in colour. Again the colour of the head region looks like a blue band from the egg shell and the egg appears to have a black spot. This is also known as (9) The ***pigment stage or eye spot stage.***At this stage the embryo is considered to be in completing stage. Along with the head region, the trunk of the embryo also becomes coloured. The entire egg appears to be having a blue band. Eggs at this stage are called (10)"blue eggs stage" and through the egg shell, the body of the embryo is found to be completely formed.

After completing the development the embryo eats up part of the egg shell around the micropole so as to come out of the egg. It is the gastric fluid or saliva that enables the embryo to rupture the egg shell. Hatching can be controlled under dark conditions and utilizing this feature hatching can be made uniform once the eggs are in blue stage.

The various organs formed from different germinal layers.

Organs Forming each Germ Layer

Germ Layer	*Organs*
Ectoderm	Simple eyes, accessory appendages, integument, sense organs trachea, silk gland, nervous system, salivary glands, malphigian tubules, external gentalia, moulting gland, oenocyte, allata body
Mesoderm	Muscles, blood corpuscle, fat tissue, genital organs, dorsal vessel, sub oesophageal gland, pro thoracic gland, circum tracheal gland pericardial cells.
Endoderm	Midgut

Development of Embryo in Common Acid Treated Eggs and Non Diapauses Eggs

Duration after–Acid Treatment	*Embryo–Development*
Acid treatment–Completion	Stage of Daruma
24 hours–1 day	Abdominal appendages appear
2 days	Stage of Constriction
3 days	Stage of pre involution
3 days & 12 hours	Stage of involution
4 days	Completion of involution
5 days	Tubercles appear
6 days	Colouration stage of trachea
7days	Stage I of Blue spot
8 days	Stage I of Blue
9 days	Hatching

Comparison of the Embryo Developing in Diapause Eggs and Non Diapause Eggs (Common acid treatment)

Eggs of non diapause variety hatch in about 10 days after ovipositon when preserved at 25 C. the formation of the guitar shaped embryo stage in the case of non diapause

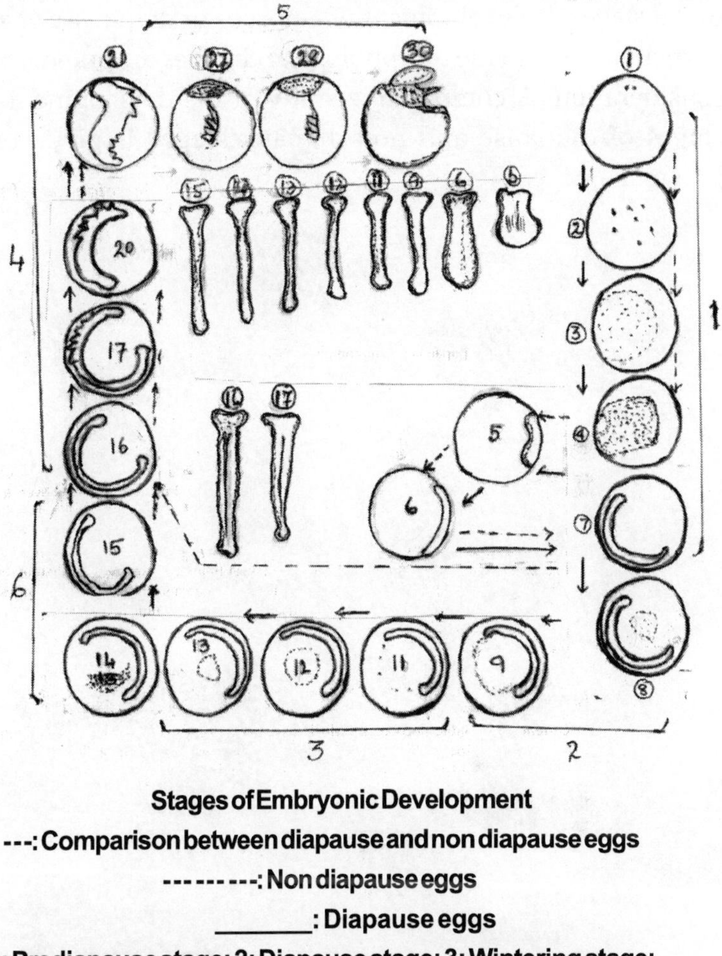

Stages of Embryonic Development

---: Comparison between diapause and non diapause eggs

--------- ---: Non diapause eggs

________: Diapause eggs

1: Prediapause stage; 2: Diapause stage; 3: Wintering stage; 4: Organs forming stage; 5: Completed stage; 6: Critical stage.

eggs and also in the common acid treatment of diapause eggs follow the same pattern as that of the diapause eggs. Later the embryos rapidly enter the organ forming stage. Although it may not be, absolutely correct, the embryo development of non diapauses eggs is found to skip some of the stages of development of embryos of diapauses eggs particularly when the morphological changes are taken into consideration. A comparative study of the developmental stages of diapause and non diapauses eggs is given on preveious page.

Chapter 16
Moultinism in Silkworm

Moulting

Literally it means to cast feather or body covering. The periodic cast of body covering at a particular interval in life cycle of an insect is known as ecdysis or moulting or apolysis. It is both genetically and hormonically controlled.

In insect during larval stage there are two phases:

1) Feeding phase and
2) Moulting phase.

Feeding Phase

At this phase larva eats voraciously

Moulting Phase

After feeding phase the skin is shed. This phase is known as moulting phase. In this phase worm's skin become loose, wrinkled and keep its mouth upward, translucent and dull and do not take any food.

Types of Moulting

There are three different kinds of moulting in silkworm such as:

i) ***Trimoulter***: moulting takes place 3 times in its larval stage ***e.g.*** Primitive silkworm.

ii) ***Tetra moulters***: moulting takes place 4 times in its larval stage ***e.g. Bombyx mori***,L. the commercially using silkworm.

iii) ***Penta moulters***: moulting takes place 5 times in its larval stage ***e.g. Bombyx mandarina.***

Genetically Point of View of Moulting

Due to the economic importance the influence of environment on moulting characters has been eagerly studied by many workers. So far, it has become known that temperature during incubation and rearing, as well as nourishment affect the manifestation of this character to a great extent.

Toyama (1912) was the first to study the inheritance of moulting. He found that trimoulting is linked to the cocoon colour gene F (flesh), with the recombination value of 11.2 per cent. Both belong to the sixth linkage group. Making use of this linkage relation, he was able to reveal the multiple allelic relation between three major genes for moulting, ***i.e.***, M^3, $+^M$ and M^5 along with some modifiers. Nakamura (1932), on the other hand, proposed a two gene hypothesis to explain the results of crosses, the dominance relations being tri> tetra> penta. His results are summarized as follows:

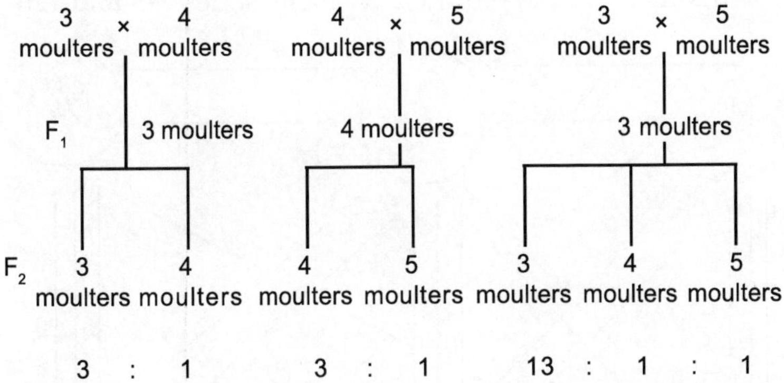

It should be noticed, however that in most cases the segregation ratios deviated more or less from those expected. Furthermore, the degree of deviation depends on the trimoulting strain used for the cross. After getting almost the same results as Nakamuras, Morohoshi advanced the view that moulting is controlled by two kinds of genes which act antagonistically and that there may be several trimoulting genes and/or strains of different potency. According to him the typical segregation ratio may be expected when a strong dominant moulting gene is involved, while segretion may be complicated if that gene is weak.

There are three sex-linked maturity genes responsible for voltinism that play important role in moultism.

1) Lm^{e}–early maturity-responsible for multivoltine
2) $+^{Lm}$–intermediate maturity–responsible for bivoltine.
3) Lm–late maturity- responsible for univoltine.

The dominant relationships of these genes are as follows: Lme>+Lm>Lm

Interaction between the two allelic series M3 and Lm

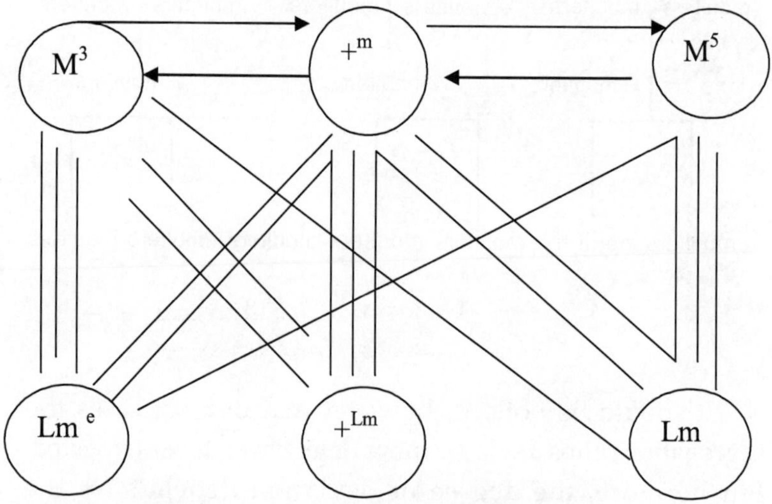

Diagramatic Representation of Interaction Between the Major Moulting Genes and Sex Linked Maturity Genes

,=,–, represents strong, intermediate and weak relations. (Morohoshi, 1950)

The expression of moulting characters is mainly under the control of the major moulting genes, M^3, $+^M$ and M^5, it is also subjected to modifications by the sex linked maturity genes, Lm^e,$+^{Lm}$ and Lm. The relation between M^3 and Lm^e,$+^M$ and +Lm and M^5 and Lm is balanced, while the balance is up set for other combinations, where Lm genes exert some influence on the expression the major genes. For instance, the late maturity gene (Lm) increases the number of moults, while on the contrary, the early maturity gene (Lm^e) tends to decrease their number.

Hormonal Studies on Moulting

Moulting in silkworms is controlled by secretions of hormone in the brain, the corpora allata and the prothoracic gland. The secretion of these hormones is controlled by the moulting genes and is also influenced by environmental conditions.

The brain in ***B.mori*** secretes brain hormone that activates the prothoracic gland to secrete ecdyson which initiate maturation or moulting. Soon after this hormone goes into action, the corpora allata secretes juvenile hormone (JH) which inhibits further maturation and perpetuates the larval characters resulting in moulting out and emergence of the next stage larva. At the time of metamorphosis the balance in effects of JH and ecdysion is in favour of metamorphosis.

At the time of moulting the content of the exuvial gland and malpighian tubule flows out between the new and old skins to help casting the exuvia. Action of prothoracic gland

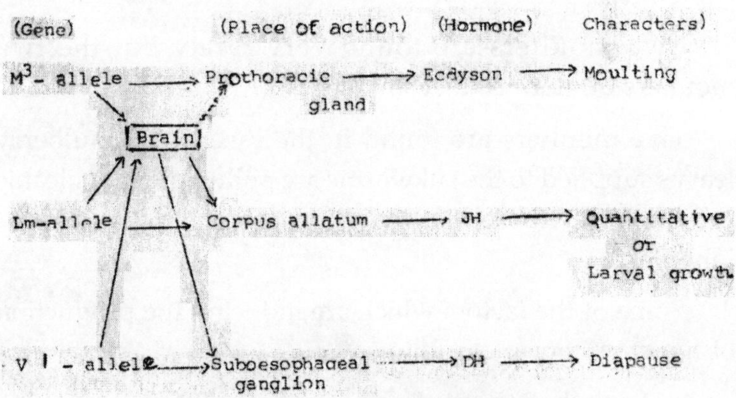

Scheme of Genetic Control of Some Physiological Characters

is controlled by the moulting genes while that of corpus allatum by sex linked maturity genes.

Environmental Factors Affecting the Moultinism

Temperature, humidity and nutritional condition affect moultinism to a great extent. The inducing factor for trimoulters is as follows

1) Low incubation temperature (20°C).
2) High temperature and humidity during the young stadia.
3) The young and soft mulberry leaves.
4) Some mulberry races are liable to produce three moulters than other races.
5) The length of light and dark times per day

Though these conditions produce the trimoulters, but the tendency of appearance of trimoulters is uncertain. Among these factors the high temperature (30°C) during the young stadia is the most effective one for producing trimoulters.

Two moulters are found very rarely but the two moulters do not produce any egg.

Five moulters are found in the case when mulberry leaves supplied to the silkworms are polluted with volcanic ash, however those five moulting characteristics are not inheritable.

Some of the factors which are inducing the production of five moulters are as follows

1) Incubation temperature is high (30°C)
2) Incubation humidity is low

3) Rearing temperature is low (16°C)
4) Rearing humidity is low (40 per cent RH)

Types of Moulting

There are atleast 5 different types of moulting in silkworm, namely di,tri,tetra,penta and hexamoulters. The most common type in the primitive domestic silkworm varieties and in the wild moths are trimoulters, where as almost all present day commercial strains are tetra moulters. Tetra molting character has been considered as the standard (normal) type in silkworm genetics.

The expression of molting character is also affected partly by the sex linked maturity genes (Lm).

Significance of Moulting Test

The significance of moulting test is to evaluate the quality of mulberry leaf. It can be considered irrespective of test breeds/hybrids,instars,rearing conditions degree of feed reduction and time of observation.

Moulting Ratio in Silkworm Under Different Conditions During First Instar

Treatment	*Moulting Ratio (per cent) With 9 feed*	*Wt. of Moulted Larvae (mg)*
Tender	78.19	4.471
Medium	46.79	3.867
Mature	31.67	3.085
Non shaded leaf	78.19	4.47
Shaded leaf	45.23	1.83
Kunva 2	78.19	4.47
Local	62.30	1.84

Chapter 17

Voltinism in Silkworm: Factors Influencing Voltinism

Voltinism refers to the number of broods raised per year. It may also referred to as the number of generations of an individual in a year. Voltinism is controlled by genes but it is also affected by environmental conditions and hormonal action and also by other genes controlling the life span of the insect, particularly the larval life.

Types of Voltinism

There are three types of voltinism in silkworm such as univoltine, bivoltine and multi or polyvoltine.

Univoltine

Univoltine races produce only one generation per year. The eggs laid remain in a diapausing condition till the next spring. Larvae of univoltine are very sensitive to temperature and other environmental conditions. They are unsuitable for summer and autumn rearing by artificial breaking of egg diapause. The larval period is very long. All European races are univoltine. The cocoons produced are commercially very superior.

Bivoltine

Bivoltine races have two generations per year, the first generation adult developing from eggs hatched in spring lay non-diapausing eggs. The second generation adults developing from these eggs lay eggs which remain in the dormant state till next spring. The larval duration is as long as univoltine. Larvae are robust and tolerate environmental fluctuations. They can be used for summer and autumn rearing and three crops can be raised per year. The cocoons are commercially superior. Japanese and Chinese races have both uni and bivoltine varieties.

Multivoltine or Polyvoltine

Multivoltine or polyvoltine have more than three generations per year. The larval duration is short and larvae are resistant to high temperature and high humidity. Larvae and cocoons are small in size. Commercially cocoons are of poor quality. The adults lay non-diapausing eggs. ***e.g.*** Nistari, Pure Mysore, Sarupat, Tamil Nadu white and Raj.

Characteristics of Different Races Based on Voltinism of *Bombyx mori*

Characteristic Feature	*Univoltine and Bivoltine*	*Multivoltine*
Egg	Diapausing/ Non Diapausing	Non Diapausing
Length of the silk filament in cocoon	1000–1600 m	300-400
Shell Ratio	15–25 per cent	10–12 per cent
Raw Silk Percentage	80–85 per cent	40–45 per cent

Factors Influencing Voltinism

When a female moth lays eggs which hibernate or which do not hibernate depends in part on the environment (particularly, temperature and light) and in part on gene differences at several loci. As far as the genetic differences are concerned, there are at present two schools of thought.

1) Nagatomo believes that Voltinism is to be determined by a series of three sex-linked multiple alleles, Hs, Hs^2 and hs (which may or may not be same as the sex-linked maturity genes Lm, $+^{lm}$ and Lm^e), modified by a number of autosomal genes, each of small effect.
2) Morohoshi: on the other hand believes the major genes determining voltinism to be a series of three multiple alleles $V,^1 +^v$ and V^3 on the sixth chromosome, although their action is affected to a great extent by sex-linked maturity genes. Perhaps there are two mechanisms involved, due to sex-linked and the sixth chromosome genes, and which mechanism predominates may depend on environmental conditions.

Toyama (1906) started experiments on Voltinism and got complicated data, as such he was unable to draw any conclusion. McCracken (1909) published her data on a cross between uni and bi voltine races; she believed that voltinism is not inherited in accordance with Mendel's rules. But from her data Castle (1910) drew conclusion that her results should be interpreted as Mendelian inheritance and that univoltine is dominant over bivoltine. As a result of his investigations Toyama (1913) described voltinism and some egg colours as maternally determined. The mode of inheritance of the former being much more complicated due to influence of incubation temperatures. He discovered that incubation of eggs of bivoltine races at 15–18 C gave rise to moths that lay non-hibernating eggs; whereas incubation at 30 C resulted in hibernating egg layers.

Since then the genetic analysis of voltinism and its responses to environment have been studied by a number of workers.

After repeated experiments, Watanabe (1918b) confirmed Toyama's findings of the influence of temperature. When bivoltine eggs were incubated at 15 C after blastokinesis of the embryo, the moths that are developed from them lays non hibernating eggs whereas when eggs were incubated at temperatures about 24 C the resulting moths laid hibernating eggs. He also found that multivoltine races were hardly affected by high temperature through his crossing experiments, he demonstrated clearly that bivoltine is dominant to multivoltine and that the case represents typical maternal determination.

Extending his experiments further to univoltine, Watanabe (1919) revealed also that univoltine is dominant to multivoltine so that segregation of 3:1 takes place in F_3.

Thus Watanabe (1924) arrived at the conclusion that uni, bi and multivoltine characters are controlled by one gene, the univoltine and bivoltine being dominant over multi voltine. He further set forth a hypothesis that hibernation depends upon the amount of an inhibitory substance present in the cytoplasm of the egg cell and that its amount is genetically controlled. However, it was not clear whether or not all genes related to inhibitory activity belong to the same multiple allelic series.

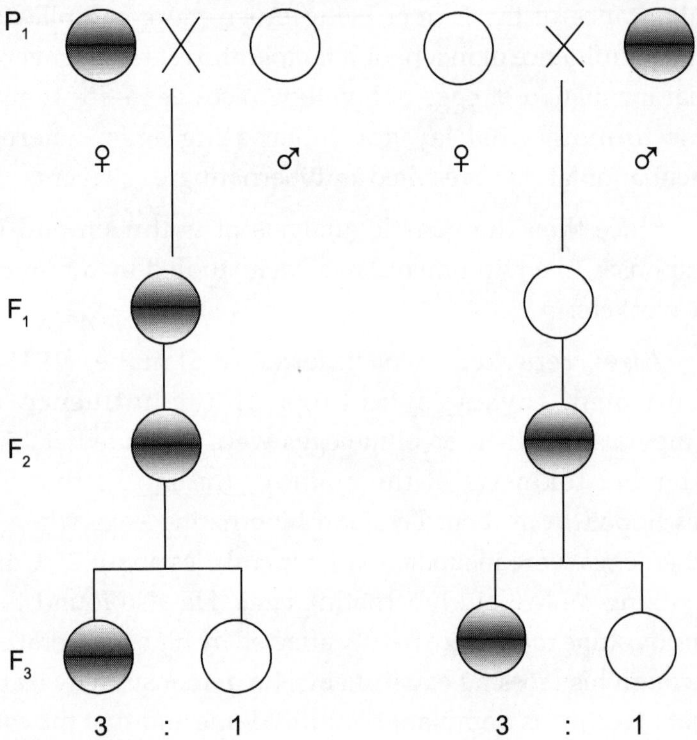

Inheritance of Voltinism, Hatched Circles Denote Univoltine and Open Circles Multivoltine

The concept of an inhibitor was followed up by Umeya (1925,1926).After ovarian transplantation and blood transfusion experiment between different races, Umeya came to the conclusion that a "Voltinism determiner" ***i.e.,*** a particular enzyme, present in body fluid, is transmitted to the egg, prediaposing it to hibernation.

In the mean time Kogure (1933) carried out extensive physiological experiments on the effect of environmental factors on voltinism. He found that light as well as temperature has a strong effect on voltinism, especially during the incubation period, long days of more than 16 hours being comparable to high temperatures and short days of less than 12 hours to low temperatures.

He also revealed that the environmental factors vary in action according to the developmental stage, the embryonic stage,especially after blastokinesis, being the most sensitive, while they are less effective in early larval stages. In the late larval and pupal stages, on the contrary, their action is reversed so that high temperature brings about non hibernating egg layers, where as low temperatures below 20 C result in hibernating egg layers.

Embryonal Stage	*Larval Stage*		*Pupal Stage*	*Character of Moths Treated*
	1st–3rd Instar	*4th–5th Instar*		
Above 25°C light 16 hours	Regardless of temperature and day length			All the moths lay hibernating eggs
18–22°C light 16 hours	Above 28°C light 16 hours	Below 20°C		Most of the moths lay hibernating eggs
18–22°C light 12 hours	Below 20°C light 12 hours	Above 28°C		Most of the moths lay hibernating eggs
Below 15°C light 16 hours	Regardless of day length and temperature			All the moths lay non-hibernating eggs

Environmental Factors

The influence of temperature on voltinism was found by Toyama (1913). He observed that incubation of eggs of bivoltine races at 15-18 C gave rise to moths that laid non hibernating eggs; where as incubation at 30 C resulted in hibernating egg layers. This finding is supported by several other scientists.

Kogure (1933) found that light as well as temperature has a strong effect on voltinism, especially during the incubation period, long days more than 16 hours, being comparable to high temperatures and short days of less than 12 hours to low temperatures.

He also revealed that the environmental factors vary in action according to the developmental stage, the embryonic stage, especially after blastokinesis being the most sensitive, while they are less effective in early larval stages. In the late larval and pupal stages, on the contrary, their action is reversed so that high temperature brings about non hibernating egg layers, where as low temperatures below 20 C result in hibernating egg layers.

Genetic Factors

There are two schools of thoughts regarding genetical affect of voltinism.

Morohoshi(1957) thought that the major gene responsible for voltinism character is a series of three multiple alleles. These are V_1,$+^v$ and V^2 affected to a great extent by sex linked maturity genes Lm,$+^{lm}$ and Lm^e.

Another school postulated by Nagatoma (1942) believe that the major gene responsible for voltinism are Hs,Hs^2 and hs located on the 6th chromosomes supported by some autosomal genes like H_1 h_1,H_2 h_2,H_3 h_3.

According to this the genotype for

Univoltine is (Hs) (Hs) H_1 H_1,H_2 H_2,H_3 H_3.

Bivoltine–(Hs^2) (Hs^2) H_1 H_1, h_2 h_2 h_3 h_3

Multivoltine–(hs) (hs) h_1 h_1 h_2 h_2 h_3 h_3

Hormonal Factors

Hasegawa (1962) succeeded in extracting the hibernation substance in crude form by treating heads of pupae and moths with ethanol and some other organic solvents. Injection of 0.25 of this substance into the abdomen of a female pupa was reported to be sufficient to transform its non hibernating egg layers into hibernating ones. Although, the chemical structure of this substance is still unknown, it is supposedly a lipid.

Prior to Hasegawa's success, there had been a series of experiments until Hasegawa (1951) and Fukuda (1951 a,b) found independently that the sub-oesophageal ganglion is the secretion centre of silkworm, where the hibernation substance is produced.

It had already been reported by Kikkawa (1943) that whether moths are ready to lay hibernating or non hibernating eggs is easily tested by Ehrlich's diazo reaction of young oocytes in the ovariole during the pupal stage. By applying this technique, Hasegawa carried out ligation or decapitation experiments with young pupae which had already been conditioned to lay hibernating eggs by incubation.

When female pupae were ligated between the sixth and seventh abdominal segment, no effect was observed on the hibernating character of eggs in the anterior part of the body. Where as ligation between the third and fourth abdominal segments at the same stage resulted in the formation of non hibernating eggs in the ovarioles of the posterior part. Decapitation of the newly pupated insects also resulted in the occurrence of moths laying non hibernating eggs regardless of whether the silkworms had already been determined to lay hibernating or non hibernating eggs by incubation. From these findings Hasegawa (1951) concluded that the secretion centre of hibernating substances is located in and around the head.

By implantation of various, organs, such as the brain, corpus allatum, salivary gland and sub oesophageal ganglion into a pupal abdomen. Hasegawa demonstrated that only the implantation of sub oesophageal ganglion affects the growing eggs in the test abdomen. Thus he finally arrived at the conclusion that the hibernation substance is secreted from the sub oesophageal ganglion.

However, that the sub-oesophageal ganglion alone cannot determine the hibernating character was pointed out by Fukuda (1952). While he was carrying out decapitation experiments, he noticed that the brain plays an important role in the determination of the hibernating character. Accordingly, he performed implantation experiments of brain and sub-oesophageal ganglion in various combinations: namely (1)brain alone (2) Sub-oesophageal ganglion alone, (3) brain plus sub-oesophageal ganglion separated from each other, and (4) Brain connected with suboesophageal ganglion. The experiments were arranged in two series. In one series (a) both the receipients and donors were of non-hibernating type. In the other series (b) the receipients were of non-hibernating and the donors were of hibernating type.

a) Implantation from the Non-hibernating Type

Where as implantation of (1) brain alone or (4) brain suboesophageal ganglion complex brought about no change in the hibernating character of the host, implantation of either (2) sub-oesophageal ganglion alone or (3) brain and sub-oesophageal ganglion together but disconnected was clearly effective, giving rise to hibernating type moths. This indicates that if the sub-oesophageal ganglion from a non hibernating type individual is separated from the brain, the former can produce hibernation effects.

b) Implantation from the Hibernating Type

Implantation of (1) brain alone was again not effective, but implantation of either (2) sub-oesophageal ganglion alone or (3) together with the brain was slightly effective in producing hibernation substances. In these tests, implantation of (4) brain–sub-oesophageal complex produced only moths laying hibernating eggs. This indicates that the activity of sub-oesophageal ganglion is controlled by the brain.

From these results Fukuda arrived at the conclusion that the brain is the actual organ that controls the activity of the suboesophageal ganglion and eventually the production of hibernation substance.According to him, the environmental conditions that affect the voltinism procedure acts directly upon the brain which as the situation requires either suppress or stimulates the activity of the sub-oesophageal ganglion.

The hormonal study of the brain has clearly demonstrated that the brain controls the function of both corpus allatum and sub-oesophageal ganglion through nerve commissures and that of the prothoracic gland via brain hormone. It is evident, therefore, that all those secretory glands are under the control of the brain which directs and integrates all their functions.

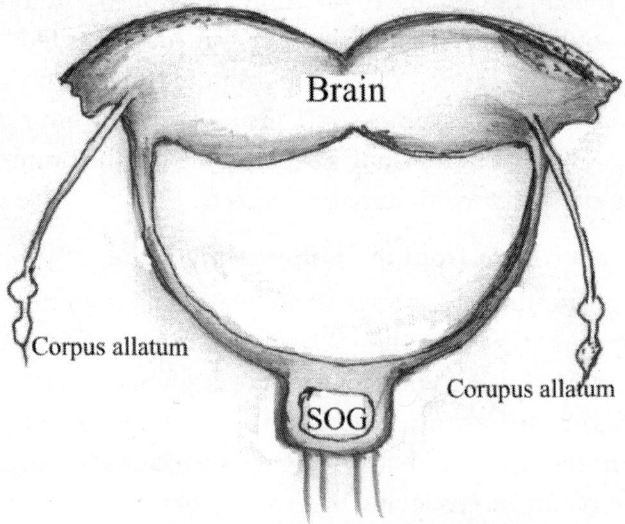

Corpora Allata and Brain-Suboesophageal Ganglion Complex of Silkworm
SOG: Suboesophageal Ganglion

Conclusion

The voltinism in silkworm is greatly affected by environment and gene although the sub-oesophageal ganglion has also effect on voltinism. Among these, which factors are indispensible are yet to be discovered.

Chapter 18
Growth

Growth in the animal body is always more or less cylindrical, period of comparative rest alternating with periods of activity. But in no group it is as evident as in insects, in which development is punctured by a series of moults or ecdyses, each preceded by a period of active growth and followed by a period in which true growth may be entirely absent.

OR

The term growth refers to many things. It may imply enlargement of a cell or of an organism, or the term may have reference to the increase in size of a population. Basically, of course growth means enlargement__ this can be brought about either by increase in size of individual cells or by multiplication of the cells.

Growth Rate

Growth in insects is generally disharmonic, heterogonic or allometric, ***i.e.***, the parts grow at rates peculiar to themselves higher or lower than the growth rate of body as a whole.

Rate of growth is mechanical primarily by moulting hormones (Wigglesworth, 1972). It is also related to the capacity of food intake, nutrients absorbed by the body from different host plants and climatic conditions.

In ***Bombyx mori L.*** growth is heterogonic and varies according to

a) Silkworm races

b) Quality and quantity of food intake

c) Climatic conditions.

Growth Ratios

Moulting is primarily a mechanism of growth, conditioned by the properties of the cuticle. In many insects the amount of growth which is achieved at each moult is predictable from certain empirical laws. It was shown by Dyar (1890), many years ago that the head capsule of caterpillars grow in geometrical progression, increasing in width at each moult by a ratio (usually about 1.4) which is constant for a given species.

The empirical laws of Dyar is:

$$Y=ak^n$$

where,

Y: Final dimension

a: Initial dimension

k: Dyar's co-efficient

n: No. of instars

Dyar's law has consequently often been used successfully to determine the number of instars in a life cycle, but there are many exceptions to it, the progression factors often changes in successive instars. Recently, it has been shown that various hemimetabolous insects (Sphodromantis, Carausius) approximately double in weight from one moult to the next,while their linear dimension increase 1.26 (*i.e.* 3□2) times. This growth factor of 2 was first demonstrated by Przibram and Megusar, it is perhaps best regarded simply to as "Przibram's factor". The physiological basis proposed by Przibram to explain his factor was that at each moult (or at each latent division) every cell in the insect body divides into two. But what histological evidence exists is definitely opposed to this improbable hypothesis. For in the Muscidae, almost the whole of the larval growth is accomplished by increase in size of the cells without multiplication, and in hemimetabolous insects there is much cellular breakdown besides the irregularly distributed cell divisions.

Growth of Tissues

There are three types of growth in tissues or organs such as

a) the increase of number of cells

b) increase in the size of cells

c) Both increase in number and size of cells.

a) Growth by Increase in Number of Cells

This is seen in the blood corpuscles, *i.e.*, though the blood is increased in volume, the size of the blood corpuscles of the infant larva is about the same as that of the grown ones. The growth of blood of silkworm is due to the increase of the number of blood corpuscles and amount of serum.

b) Growth of the Size of Cells

The growth of the silk gland exuvial gland, salivary gland, malpighian vessel, ostium valves of the dorsal vessel is due to the increase of their cell size. In these organs, cells do not increase in number and only the cell size increases remarkably. As the example of this growth there is the female sexual cells, that is the egg cells increase extremely in their volume, becoming large enough to be seen through the naked eye just before the moth emergence, while the number of cells in the ovary decreases gradually, being consumed as the nutrients for the egg cells or in the case of cells of the middle division of the silk gland, the size of the gland cell also become so big in the mature larva that they can be seen by the naked eye.

c) The Third Type

The third type of the growth is that of the fatty tissue, integument (skin) tissue or the epidermis of several organs. In the cells of those tissues (or organs) the size and number of cells of the grown larva are larger than those of the infant larva, showing the growth of the tissues in both ways, towards increase in number and volume of cells.

Growth Curve of Silkworm Larvae

The growth speed of silkworm larvae varies with instars or different time of the same instar. Usually it is slow at the beginning. At the middle age it accelerater and finally it slows down again. If we investigate the increase of body weight of silkworm larvae from 1st to 5th instar or daily increase of body weight during any instar, we can get a 'S'–shaped curve. This indicates that, the newly moulted larvae have the least body weight; after the 1st feeding, through light eating stage and moderately eating stage, the body increase rapidly, during the premoulting stage, it slows down again.

The change of growth rate can be described by mathematical formula.

According to Robertson (1909) the velocity of the growth in the weight and dimension of the living matters is slow at the beginning of the growth, largest at the middle period and later it decreases, and this phenomenon is explained by the equation of monomolecular autocatalytic reaction(single molecular chemical reaction) formula to describe the biological growth speed.

$$\text{Log } x/A\text{-}x = k\ (t\text{-}t_1)$$

where,

A: Maximum growth value

X: Actual value at the time

K: Velocity of co-efficient

t: Time measured

t_1: Time at half (1/2) value.

Yagi (1926) also explained the growth of the weight of silkworms in the 5th stadium applying the Robertson's equation. After this many investigators surveyed and discussed this. This formula is suitable for the calculation of growth curve of ovary during pupal stage, but it is not suitable for the daily change of body weight or weight of the silk gland. Chinese scholar Wu Zaide, after theoretical study on growth curve of the silkworms, suggested the following formula

$$In(X) = In(X_o) + (k_1 - K_2)t$$

where,

X: Body weight at t time

X_o: Body weight when t = 0

K_1: The maximum growth speed at an unit time.

K_2: Decreasing growth speed.

=: $[K_1-(Inx_a-Inx_0)/ta.\ 1/ta$

X_a: Body weight at maximal growth

T_a: Time of maximal growth

Effect of Environmental Condition on Silkworm Growth

The rate at which the silkworm grows is influenced by temperature which permit growth, that is higher than 10°C and lower than 28°C, development generally proceeds more rapidly at higher temperature. Humidity also affects the rate of development. The ranges of optimum temperature and humidity for growth are 26-27°C and 80-90 per cent at the 1st to the 2nd instar, 25°C, 70-80 per cent at the 3rd instar and 23-24°C and 60-75 per cent at the 4th to 5th instar.

Effect of Hormones on Growth of Silkworm

The larval growth is affected by the moulting hormone. At larval moult the effect of this hormone is modulated by the juvenile hormone, when the 5th instar larvae are administered a synthetic juvenile hormone, pupation is delayed and their body size exceeds that of the non-treated control.

Growth Compensation

Under normal condition, the growth of internal organs or tissues of the silkworms is harmonius. But if one side of the paired organs ***e.g.*** silkglands or ovaries of the 5th instar larvae is excised the another side will grow abnormal. And also, if the ovaries are excised before the 5th instar, the oviducts and accessory glands of emerged females enlarged much more than that of those females not excised. The phenomenon that the excision of one organ leads to the abnormal growth of another organ is called growth compensation.

Individually, the growth compensation phenomenon of the silkworms is also expressed in the duration of development. For example, if the eggs (after the embryonic reversal stage) and young silkworms are kept at 18°C, the young silkworm stage is prolonged and the grown silkworm stage is relatively shortened. This results in the smaller mature silkworm and less spinning. On the contrary if the young silkworm kept at 25°C–the young silkworm stage is shortened and the grown silkworm stage is relatively prolonged. This results in eating more mulberry leaves, larger mature silkworm and more spinning. The phenomenon that the shortening of the earlier stage leads to the prolongation of the later stage or by contraries is called duration compensation.

Chapter 19
Application of Hormones in Sericulture

The research and application of insect hormone opens up a new route for regulating the growth and development of silkworm and increasing silk production. Since 1936, when French entomologist Bounhiol proved the secretory function of corpus allatum in the silkworm,the research on insect hormone has made a great progress incessantly and achieved new breakthrough. Since 1950, the progress in isolation, purification, structure identification and artificial synthesis of insect hormone is very rapid. Butenandt and Karlson (1954) extracted pure ecdysone from silkworm pupae. In 1965, Huber and Hoppe *et al.* identified its structure, Nakanisi (1966) extracted phytoecdysone from plant. Williams (1956) extracted juvenile hormone (JH) from cecropine silkworm, ***Hyalophora cecropia*** L. afterwards a lot of juvenile hormone (JH) analogue was synthesized artificially one after another. These research achievements laid a solid foundation for the application of insect JH, AJH and MH in production.

1. Application of Juvenile Hormone in Sericulture

Juvenile hormone is secreted by corpus allatum and its major physiological function is to maintain the larval characteristics, since 3rd day of 5th instar, its secretion amount reduces gradually and disappears finally. Therefore, during the time when its secretion amount reduces, if we externally add some JHA into body, it can maintain larval characteristics, prolong the duration of leaf ingest and silk protein synthesis and thereby increase cocoon production.

Type of Juvenile Hormone

Since the structure of natural juvenile hormone was defined, great amount of JHA possessing high bioactivity has been synthesized. In order to screen out very effective JHA, a lot of sample were apparaised, among them,J002, J003, 738,73411,ZR515,ZR512,ZR619,ZR,ZR520,Zhengsiso and Zhengsileng etc. are highly bioactive to silkworm.

Technique of JHA Application

a) Dosage

Due to the high bioactivity, its dosage is very low; otherwise it will cause the occurrence of non cocooning silkworm. When we apply JHA and decide its dosage, we must consider the following points

1) It will not affect silkworms health and cause the non cocooning silkworm
2) It prolongs the duration of 5th instar by about 1 day.

It has a noticeable silk increasing result while it does not lower the silk quality.

Some JHA Sample and Applying Dosage

JHA	*Optimum Dosage (μg/larva)*	*Safe Dosage (μg/larva)*
J003	1.5	1–2
T34	0.1	0.05–0.5
738	1.5	1–2
Zhengsiso	1	0.5–2
Zhengsileng	1	0.5–2

Applying Period

Regulation of silkworm growth and development by JHA has a relatively strict timeliness. In order to master the suitable application time for achieving an ideal effect, JHA was sprayed once per 12 hours onto silkworm in whole larval stage. The result showed that there was no effect of duration prolonging or shortening at 1st to 4th instar except that a slight duration prolongation (1-5 hrs) when treated in the middle stage of 4th instar.

Result of body spraying to the 5th instar silkworms is completely different from that of the 1st–4th instar ones. It showed an extremely significant effect of duration prolongation.

If four time intervals were divided in the 5th instar, then, body spraying of JHA at any time in the first 3 intervals achieved an obvious effect of prolonging duration and increasing cocoon filament amount. The best application time is at the middle stage of 5th instar.

Applying Direction

Most of JHA are lipid soluble compound, so we apply JHA by body spraying. The optimum time for body spraying is when silkworms eat up leaves and expose their body, 10–20 minutes after spraying, feeding can be carried out normally.

Results of JHA Application and its Effect on Silk Quality

Increase in Cocoon Character

Within the range of a safe dose, JHA body spraying once in middle stages of 5th instar prolongs the duration by 12 to 24 hours and achieves an obvious effect of increasing cocoons characters. The result of investigation on 279 test plots showed that

a) ***Cocoon weight***: 99.4 per cent of the test plots showed higher cocoon weight than the check (control). That of 89.8 per cent of the test plots increased by more than 5 per cent and that of 53.6 per cent of the test plots increased by more than 10 per cent.

b) ***Cocoon shell weight***: All of the 279 test plots (times) were higher than the check. 96.5 per cent of the test plots increased by more than 5 per cent and 77.6 per cent of them increased by more than 10 per cent.

c) ***Cocoon shell rate***: 86.1 per cent of the test plots gave higher result than the control and most of them increased by 2-8 per cent.

Comparative test of cocoon production, which was conducted in countryside by 78 times, showed that cocoon yield per box of eggs all achieved a level of extreme significance in statistics.Cocoon reeling and silk quality identification showed no significance in statistics in cocoon reelability (92 times Test), neatness of raw silk (72 times) and raw silk percentage of cocoons (50 times test), etc.

For discovering ways to further increase in cocoon yield with JHA body spraying, many application combinations were conducted in 4th instar and 5th instar. The result showed that cocoon yield increase could exceed 20 per cent with JHA body spraying two times (ZR515,1-2μg/larvae) in the middle stage of 5th instar.

As an auxiliary technology while there was a shortage of high yielding spring race

In China, commercial silkworm eggs are divided into spring rearing and summer or autumn rearing races. In general, the cocoon productions of silkworm race for spring rearing are higher than summer or autumn rearing. At practical production, people have to utilize the silkworm race for summer or autumn rearing in spring because of the shortage of silkworm race for spring rearing or the air pollution. In this case, they can apply this technology with JHA and make the cocoon production more beyond the level of silkworm race for spring rearing.

Effect of JHA on Increase of Cocoon Filament Size

5–10 per cent coarse size cocoon filament was obtained when JHA was applied on the surface of the 5th instar larva.

Application of Moulting Hormone in Sericulture

Moulting hormone was mainly purified from plants which is known as phyto ecdysone with fairly high physiological activity. It has been extracted from number of plants such as ***Ajuga decumbens. Achyranthes aspara and Cyanotids arachnoidea.***

The Regulation Effect of MH on Silkworm Growth and Development

Investigation on the effect of duration prolonging or shortening of MH administration once per 12 hours from 1st instar to mature larva shows a complicated situation. MH administration to newly exuviated silkworms from 1st to 4th instar prolongs duration of the same instar but when administrated at primary or middle stage of the instar, the duration of the instar shortened. The effect of duration shortening slowed down gradually as MH feeding was conducted later within the instar. The effect of MH administration in 5th instar was similar to that at 1st instar to 4th instar. It is different only when the duration of the instar is prolonged; when MH was fed at the primary stage of 5th instar (about the 2nd day). So from these two useful conclusions can be drawn that

a) Silkworm at the primary stage of 2nd instar or 3rd instar is sensitive to MH. It is suitable time for identifying the activity of MH which saves labour, time and the chemical.

b) In special situation needed to shorten the duration of the instar in rearing silkworm, one can attain the goal by administrating MH at any time after the middle stage of 5th instar. The duration could be shortened by 2-3 days when MH administration has been done at the middle stage of 5th instar.

The Uses of MH Application in Sericulture

Promote the Uniformity of Silkworm Maturing

In silkworm rearing 10 per cent of labour is required at the stage of ripening of worms, for collecting the mature worms and put them on mountages and feeding the unripe worms. Particularly some of the races when confronted with the weather of low temperature are not uniformly mature. It could spend 2-3 days from earlier maturing to end maturing. So it is very difficult in mounting, mounting management, cocoon harvest and cocoon marketing. It also influences the cocoon quality. If MH is administered when the silkworm start to be mature, the maturing will be uniform. It is more useful for improving the efficiency of mounting, saving mulberry leaf, improving the uniformity of pupation and being beneficial to improve the cocoon quality.

Shorten the Duration of 5th Instar

During silkworm rearing, there is often a situation of lacking mulberry leaves at the middle or later stage of 5th instar since sometimes rearing amount was not well planed or weather is unfavorable for mulberry growing. Thus MH may be administrated to promote silkworm maturing and the problem of mulberry shortage may be overcome. At the later stage of 5th instar it could also be administrated with MH for shortening the duration and reducing losses when a severe silkworm disease was occurring.

Prevent or Reduce the Occurrence of Non-cocooning Silkworms

Non-cocooning silkworms are often found in treatment with high dose J.H. Administration of MH could reduce losses caused by non-cocooning silkworms. It was observed that the non-cocooning silkworm rate was 22.2 per cent if body spraying of J003, which is a juvenile hormone analogue was done at the rate of 2.5μg per larva at 72 hours of 5th instar, and the non-cocooning silkworm rate was reduced to 9.6 per cent after MH was administered at earlier maturing.On the other hand, MH administration could reduce occurrence of non-cocooning silkworms caused by high temperature or rearing with tender leaves.

In addition, administration of MH at the earlier stage of 5th instar can increase cocoon filament yield by various degrees.

Effect of MH Application on Silk Quality

There is no bad effect on silk quality ***i.e.***, reelability, filament length, wt. of filament and neatness on the application of MH.

Application of Anti-juvenile Hormone in Sericulture

Anti-juvenile hormones are not actually the insect hormones. It is an antihormonal agent isolated from plants on the basis of natural defensive mechanism of the plant from insects. Its chemical structure was identified and named as precocene. Precocene means precocious metamorphosis. The available antihormonal agent in the market which has been tested on silkworm for the development of trimoulter are SM–1, Lin–II. YA20, CA1, Ca2, CA3 and CA4. The level of trimoulter's induction rate depends not only on the hormone activity, healthiness of the silkworm and the amount of ingested food, but also on the instar in which the hormone is applied. The degree of easiness of inducing trimoulters with AJH was 4th instar>3rd instar>2nd instar>1st instar. Within the 3rd or 4th instar, administration of AJH at the early or middle stage of the instar showed a similar induction rate of trimoulter in silkworms, ***i.e.***, it was strictly timely. There was no moultinism changing by administration of AJH at 5th instar, on the contrary, the duration of the instar was slightly prolonged and the cocoon wt. was increased.

The most advantage of application of AJH in sericulture is that it can control the cocoon filament size. We could get 2–3 kinds of cocoon filament size in the same race by applying AJH and could get the raw silk of high quality materials for developing new products of high quality and thin silk fabric, and for reeling silk wrapped by fine size silk.The cocoons obtained by administration of AJH at 3rd or 4th instar have very good reelability, good neatness, little difference of cocoon filament size and well improved raw silk strength and stretching ability.

Chapter 20
Artificial Diet of Silkworms

The first artificial diet for insect was developed by E.Bogdanow in 1908. He reared Blow fly ***Calliphora vomitoria*** from egg to adult on a diet of meat extract, starch, peptone and mineral salts. Bottger (1942) and Beck (1949) developed diets using highly purified natural products for Euporean corn borer, ***Pyrausta nubilalis***. These two works present the beginning of the development of modern artificial diets for phytophagous insects.

The concept of formulating an artificial diet for silkworm, ***B.mori*** emerged principally from the curiosity and eagerness of a Japanese sericulture professor, Yasuji Hamamura in 1935. He put forth a highly pertinent question "why silkworms feed exclusively on mulberry leaf? ". He extracted a bio- chemical principle from mulberry leaf in alcohol and absorbed it into a filter paper. The filter paper started smelling like mulberry leaf and the brownish wilted residue did not have the original smell. He placed the filter paper and the brownish leaf residue at the opposite ends of a box and 10 silkworm larvae were placed in the middle of the box. The box was closed for 30 minutes and when opened, it was found that odour absorbed into the filter paper attracted silkworm to the filter paper. But the larvae collected on the filter paper never showed any sign of biting the filter paper. They raised their heads and shook them right and left. But when the larvae were placed on the mulberry residue from which the odour was removed, it was seen that the larvae started eating the residue vigorously. The methanol extraction of the residue was subsequently carried out; the extract was added to agar agar. The larvae placed on this agar agar began biting it. By this it became clear that odour of mulberry leaf only attracts the silkworm and they eat it due to some stimulus other than the odour. In 1948, Torii and Mori reported that one of the maxillae of silkworm is sensitive to the odour of mulberry leaf and the other is associated with the feeding behaviour.

Hamamura pointed out that the biting of silkworm did not continue long when the methanol of fresh mulberry mixed with agar agar was provided to silkworm but it continued when the residual mulberry was also added to it. This again gave an idea that a substance, which assists the continuous swallowing, is separately present in the leaf and the biting factor and the factor, which assists continuous swallowing, are different from each other. So, the relationship between mulberry leaf and silkworm was elucidated that the leaf contains an attracting factor which is sensed by the sensory cells in the maxillae. Thus it became clear that any artificial formulation should basically contain the above said three factors.

The terpenes citral, linalyl acetate, linalool and terpinyl acetate found in the mulberry are the main attracting factors (Hamamura and Naito, 1960). In fact silkworms are attracted by both fresh leaf alcohol and fresh leaf aldehyde which are the typical components of green leaves (Watanabe, 1958). The attraction tests of these 4 pure chemicals indicated that the degree of attraction varies among them. It was also reported that besides these four attractants, there are several other terpenes and esters which also have the same function (Hirao and Ishikawa, 1964).

There are two types of biting factors, the ether soluble □-sitosterol (Hamamura ***et al***, 1961) and the water soluble isoquercitrin (isoquercitrin) of the flavon group (Hamamura et al, 1962,a). the pigment morin found in mulberry leaf also has the same function. □-sistosterol is present in the waxy material covering the surface of mulberry leaves and provides the stimulus to the maxillae of the silkworm. Isoquercitrin is present in the cells of the leaves and is believed to assist the action of □-sitosterol.

Swallowing factor present in the mulberry leaf is the cellulose which induces swallowing in silkworm (Hamamura ***et al.***, 1962, b). However, cellulose alone is not sufficient for this action. The presence of potassium phosphate is essential to reinforce the action. Sugar or inositol is essential to reinforce the action. Sugar or inositol is essential to obtain complete swallowing action. Cellulose is also believed to stimulate the peristaltic movement of the intestine of the silkworm and expel the faeces smoothly.

Olfactory and the gustatory organs known as chemo receptors are involved in food selection in silkworm and decide whether the food is suitable for consumption or not. Olfactory organs located on the antannae sense the taste or odour of the food and maxillary palpi make the larvae to get attracted to a particular food. When the larvae come in contact with food, the odour is sensed by receptor cells present on the antannae and maxillary palpi and is charged to eliminate the feeding inhibition and induce the larvae to initiate feeding. Whether feeding is continuous or not lies on the gustatory organs located on the maxillary tubercle body (Ishikawa, 1967). If the odour of food can activate sugar receptor cells of maxillary tubercle and if there are no repellent substances, which stimulate picric receptor cells, the larvae continue feeding. The preference of food is totally controlled by the sensory and gustatory organs of the silkworm. These findings enabled formulation of artificial diet for silkworm.

Developing an artificial diet implies challenging the natural mechanism of feed selection by silkworm and continuance of feeding, which involves blending all those constituents that are present in mulberry leaf in dry form. Such an artificial diet should meet the nutritional requirements of silkworm. It should have the proper feeding inducers. The physical properties of the diet such as the texture and moisture content should be such that silkworm feeds on it continuously as done on mulberry leaf. It should be stable and antimicrobial.

Fukuda (1960) started working on formulating a viable artificial diet and he was the first to rear silkworm successfully on artificial diet from I to V instar. The silkworms reared on artificial diet showed retarded larval growth and development, small cocoon size and reduced fecundity.

Composition of Artificial Diet Developed by Fukuda (1960)

Substance	*Content(g)*
Mulberry powder	5.0
Starch	1.5
Sucrose	1.0
Quina powder	1.0
Defated Du Fu	1.0
Anticeptics	Added
Distilled water	15 ml

The dietary composition of artificial diet has been improved continuously. Now hundreds of diet for various purposes has been developed. And rearing silkworms, especially the young ones on artificial diets become more and more popular in Japan and in some places of China.

The application of artificial diet in sericulture has many advantages, such as

1. The silkworm can be reared throughout the year.
2. Labour can be saved
3. Disease infection can be reduced

The main disadvantage of rearing silkworm with artficial diet is that the prices of artificial diet are still higher than those of fresh mulberry leaves. So, it is necessary to develop cheaper diet in future for research works. It is wise to rear silkworm on artificial diet up to 3rd stage and the subsequent instars on fresh mulberry leaves.

Types of Diet

The artificial diets can be divided into three basic groups.

Oligidic Diets

Oligidic diets consist principally of crude natural materials such as mulberry leaf powder. These types of diets are used in mass-rearing projects or for studies that do not involve nutrition as components are readily available and inexpensive. The cocoons are usually produced in mass and used in studies that do not involve nutrition.

Oligidic diets are popular because of the general simplicity in preparation and involvement of low cost. The composition of some of these types of artificial diet is given below.

Composition of Some Oligidic Diets for Young Silkworm

Substance (g)	*Ito (1968)*	*Zhang (1992)*	*Chen (1994)*
Mulberry leaf powder	50	30	35
Defeated soya meal	20	35	35
Potato starch	20	–	–
Corn starch	–	11.5	11
Soya residue		15	–
Sucrose	–	–	–
Cellulose	8	–	10
Ascorbic acid	2	1.5	1.5
Citric acid	0.5	2	2
□–sitosterol	–	0.4	0.4
Salt mixture	–	3	3
Vitamin B mixture	added	1.0	1.5
Agar	12	–	–
K_2HPO_4	1	–	–
Antiseptic	Added	Added	added

Meridic Diets or Semi-Synthetic Diets

They are composed mostly of defined chemicals but with one or more ill-defined components such as defatted soya meal.

The nutritional studies with artificial diets, both qualitative and quantitative contributed a great deal of information regarding silkworm dietries. These are:

1) Starter (for the first instar larvae),
2) Grower (for the 2nd to 4th instar larvae),
3) Producer (for the 5th instar larvae).

In order to make these diets more palatable, pulverized dried mulbery leaves or some leaf components were added to the diet 10 to 20 per cent of mulberry leaf powder on the dry weight basis is added for younger larvae to accelerate their feeding. In course of time the diet without leaf powder were used for the IVth and Vth instar larvae.

Composition of Meridic or Semi-Synthetic Diets Without Mulberry Leaf Powder (Ito, 1974)

Substance	*Dry Wt. (gms)*		
	1st Instar	*II-IVth Instar*	*Vth Instar*
Potato/Starch	10.0	10.0	20.0
Sucrose	10.0		
Glucose	–	–	12.0
Soyabean meal defatted	30.0	40.0	60.0
Soyabean oil(refined)	3.0	3.0	3.0
□-sitosterol	0.5	3.5	2.0
Salt mixture Wasson's	3.5	3.5	2.0
K_2HPO_4	1.0	1.0	–

Contd...

Contd...

Substance	*Dry Wt. (gms)*		
	1st Instar	*II-IV^{h} Instar*	*Vth Instar*
Ascorbic acid	2.0	2.0	2.0
Cellulose powder	34.0	34.0	–
Agar	15.0	15.0	5.0
Citric acid	0.5	0.5	0.5
Sorbic acid	0.2	0.2	0.2
Morin	0.2	0.1	–
Total	109.9	119.8	105.2
Vitamin B mixture	Added	Added	Added
Distilled water	300ml	300ml	260ml

Holidic Diet

These diets consist entirely of known pure chemicals. But in strict sense a holidic diet for silkworm may not be possible as contaminants are commonly present in many of these components such as (agar) used in the diet. Perhaps the amino acid diet may be regarded as one of the holidic diets. The holidic diets are most desirable for critical nutritional studies.

Composition of Amino Acid (Holidic) Diet (Ito *et al.*, 1966)

Substance (gms)	*Quantity*
Potato/Starch	5.0
Sucrose	15.0
Amino acid	20.0
Refined soyabean oil	3.0
β-sitosterol	0.5
Wesson's salt mixture	4.0
Ascorbic acid	2.0

Contd...

Contd...

Substance (gms)	*Quantity*
Cellulose powder	35.2
Morin	0.3
Agar	15.0
Total	100.0
Mulberry leaf	10mg/g
Biotin	4μg/g
Cholin chloride	1500μg/g
Folic acid	4μg/g
Inositol	400μg/g
Nicotinic acid	300μg/g
Pyridoxine	30μg/g
Riboflavin	20μg/g
Thiamine	20μg/g
Sorbic acid	Added

The detail composition of artificial diet used for practical purposes are as follows:

Substance	*Diet for 1st-IVth Instar (g)*	*Diet for Vth Instar (g)*
Mulberry leaf powder	25.0	25.0
Soybean oil	1.5	3.0
Defatted soybean meal	36.0	45.0
Cholesterol	0.2	0.2
Citric acid	4.0	4.0
Ascorbic acid	2.0	2.0
Sorbic acid	0.2	0.2
Agar	7.5	5.0
Salt mixture	3.0	3.0

Contd...

Contd...

Substance	Diet for 1st-IVth Instar (g)	Diet for Vth Instar (g)
Glucose	8.0	10.0
Potato starch	7.5	15.0
Cellulose powder	20.8	-
Vitamin B mixture	Added	Added
Antiseptic	Added	Added
Total	1/5.7	1/2.4
Water	300ml	300ml

Physical properties of diet

One of the important aspects of diet improvement is the physical properties of the diet. It includes the shape, textile hardness, viscosity and uniformity, which have direct effects on the feeding activity, growth and development of larvae. The water retention capacity of the diet was improved by addition of more moulding materials like starch, agar, glue etc. The hardness of prepared diet was found to influence the dietary efficiency greatly. Cellulose powder when added to the diet was found to have stimulating effect on feeding as water content influence the dietary efficiency. About 75 per cent water content was found to be most favourable so far efficiency and food consumption is concerned.

However, suitable water content must be divided separately for each diet because the basal composition differs in every diet.

Antibiotics and antiseptics were usually added to the diet to prevent decomposition.

Basic criteria for an artificial diet

An artificial diet has to meet four basic requirements.

a) Essential nutrients like proteins, carbohydrates, lipids etc.

b) Solidifying chemicals,like cellulose,agar etc

c) An attractant substance like citric acid or □-sitosterol

d) An antiseptic substance, like ascorbic acid, chlormycetin and benzoic acid (to make it free from all microbial contamination).

Preparation

1) The components of the diets are ground weighed and mixed thoroughly. Then it is transferred with water containing vitamins and antiseptic into a container covered with a lid.
2) Then the whole mixture is heated to approximately 97 C on a water bath for 15-20 minutes and then cooled to form a gel.
3) Diets can be stored in refrigerator for few days.

Mode of Application

The prepared diet is in the form of gel and it is cut into thin slices and fed to the larvae once a day or on every alternate day.

Prospects of Artificial Diet

Practical use of artificial diet in sericulture is thought to be economically very desirable. Large scale rearing could be initiated with artificial diet feeding the larvae upto 3rd stage and subsequent instars on fresh mulberry leaf.

Inorganic Salt Mixture (gms)		Vitamin Groups (gms)	
$Ca CO_3$	5.0	Thiamine	0.020
$Ca_3 (PO_4)_2$	0.4	Riboflavin	0.020
$K_2 HPO_4$	10.0	Pyridoxine	0.030
$Fe PO_4.H_2 O$	1.9	Nicotinic acid	0.100
$MnSO.5H_2 O$	0.05	Pantothenic acid	0.150
$ZnSO_4.7H_2 O$	0.30	Biotin	0.020
$MgSO_4.7H_2 O$	0.30	Cholinechloride	1.500
Corn starch	19.55	Inositol	2.000
Total	40.00	Folic acid	0.002

Position of Artificial Diet of Silkworm in India

In the global market silk produced in India has not been up to the desired level. Although the recent efforts in evolving and releasing highly productive and elite silkworm breeds have trend setters in improving the quality of silk yarn the laurels gathered galore have not contributed to an absolute face lift to the Indian produce yet. One of the reasons for slow transformation is that still, more than 85 per cent of the Indian sericulture is banking on multi X bivoltine silkworm seeds for commercial exploitation inspite of the high impetus given to the bivoltine sericulture. The Indian mulberry by and large suffers an inconsistent nutrient level particularly the one offered to young instar silkworm. It has been further more difficult to improve the quality of the mulberry leaf available in tropical India.

It is widely known that a better young instar leads to a healthy cocoon crop which further leads to quality produced. For an assured and quality cocoon crop, the balanced nutrition and desired moisture are the prerequisites principally during the young instar. Keeping this in view, sustained efforts were made by the CSR and TI, Mysore to formulate an artificial diet for young stage silkworm. The premier research funding agency, the department of Biotechnology has assisted the programme initially understanding the need and potential of the initiative. The idea was to formulate a viable artificial diet for the young instar (upto the second moult) and then shift it over to mulberry leaf. The whole constraint what India faced and is facing now are totally different from that of Japan-the pioneer of artificial diet technology-faced. Japan had to overcome the fast decline in availability of land for mulberry production and to facilitate near complete automation of sericulture.

The artificial diet developed in India was not particularly well in respect of the response of silkworm to the diet. When compared to the Japanese diet and mulberry leaf the performance was below par with respect to the weight gain and the duration (Magadum ***etal*** 1994). Although the bivoltine pure breeds showed good response initially, the same was not sustained. Efforts were also made by other groups to develop artificial diet especially for bivoltine sikworm breeds and hybrids (Hareesh Kumar and Thakur, 1999) but as the result were not consistent enough; the same were not translated true to the content and sprit, let alone its field adoption.

The work on artificial diet had been started in 1992 at CSR and TI Berhampore (WB) and a diet for late age has been developed but the diet for young stage was not good enough, only 60 per cent survivality was achieved. Due to lack of financial crisis, the authority of CSB suggested to drop the programme at CSR and TI Berhampore but the same work is going on in other institutes.

The sustained efforts at CSR and TI,Mysore, to develop the diet for silkworm gave fruitful results towards the mid and late 1990s. The inexpensive diet developed using cheap ingredients was well accepted by the popular multi x bivoltine hybrid,PMXMB$_4$ D$_2$ (Magadum and Shivakumar,1995). The feeding response ranged between 88 and 96 per cent among the various formulations and the larval weight gain was reasonably fair although comparison with mulberry leaf fed batches was not very encouraging. The improved feeding response was of course a silver lining on one hand but a major bottleneck on the other hand because this response was largely hybrid specific. Although this gave hopes and chances for further improvement of the diet composition and various related techniques to improve upon the very appearance of the diet, the big challenge was looming large to make the diet equally acceptable to other productive and elite breeds and hybrids.

Efforts to develop artificial diet for tasar silkworm ***Antheraea mylitta*** (Patnaik ***etal***,1985) and for eri silkworm, ***Philsamia ricini*** (Choudhury,2000) were not successful. The tasar and eri efforts were mainly to boost laboratory research interest and to do the young stage rearing in ***Bombyx*** way although the formulation had obvious modification in the composition.

References

Altman, G. (1956) Die Regulation des wasserhaltes der Honigbiene. Insects Sociaux, 3: 33–40.

Beck, S. D., Lilly, J. H., Stauffer, J. F(1949) Nutrition of the European cornborer, ***Pyrausta nubilalis***(Hubn). I. Development of satisfactory purified diet for larval growth. Ann. Entomol. Soc. Am. 42:483-96.

Bottger, G. T. (1942) Development of synthetic food media for use in nutrition studies of the European corn borer. J. Agr. Res. 65:493-500.

Bounhiol, J. J. (1936). Metamorphose apres ablation des corpora allata chez la ver a soic (Bombyx mori) C. R. Acad. Sci. 203: 388.

Craig, R. (1960). The physiology of excretion in the insect ***in*** Ann. Rev. of Entomology, (Vol-5):53-68.

Castle, W. E. (1910) The effect of selection upon Mendelian characters manifested in one sex only. J. Exp. Zool., 8: 185-192.

Choudhuri, C. C. (2000). Rearing of eri silkworm, ***Philosamia ricini*** and ***Philosamia cynthia ricini*** on artificial diet. Abstr. National conference on strategies for Sericulture Research and Development, November, Mysore, India, p. 94

Chapman, R. F. (1998). The Insects(4th Edition)–Cambridge University Press.

Dyar, H. G. (1890). The number of moults of lepidopterous larvae. Psyche, 5: 420-422.

Fukuda, S. (1951a). Factors determining the production of the nondiapause eggs in the silkworm. Proc. Jap. Acad., 27:582-586

Fukuda, S. (1951b). The production of the diapauses eggs by transplating the suboesophageal ganglion in the silkworm. ***Ibid***., 27:672-677.

Fukuda, S. (1952). Function of the pupal brain and sub–oesophageal ganglion in the production of the non-diapause and diapauses eggs in the silkworm. Annot. Zool. Jap., 25:149-155.

Fukuda, T. (1960). Rearing of silkworm on artificial diet. Agriculture and Horticulture, 35:103-104.

Hiratsuka, E. (1917). Researches on the nutrition of the silkworm. Bull. Sericult. Exp. Sta., 2:353-412.

Hiratsuka, E. (1920). E. Bull. Imp. Sericult. Expt. Sta. Japan, I:137-315.

Herbert H. Ross (1956). A text Book of Entomology (2nd Edition)-John Wiley and Sons, INC, NewYork and Chapman and Hall Ltd., Londal.

Hamamura, Y. and Naito, K. (1960). Food selection by silkworm larvae, citril, linalool as attractants of the larvae. Nature, 190:879.

Hamamura, Y., Hayashiya. K., and Naito, K. (1961). Food selection by silkworm larvae, B-sitosterol as one of the biting factors. Nature, 190:881.

Hamamura, Y., Hayashiya, K., Naito, K., Matsura, K., and Nishida, J. (1962). Food selection by silkworm larvae, water soluble biting factors. Nature, 194:754-755.

Hasegawa, K. (1951). Studies on the voltinism in the silkworm, ***Bombyx mori*** L., with special reference to the organs concerning determination of voltinism. (A preliminary note). Proc. Jap. Acad., 27:667-671.

Hasegawa, K. (1962). On the action of diapause hormone in the silkworm. J. Seric. Sci. Jap., 31:213.

Hamamura, Y., Hayashiya, K., Naito, K., and Matsura, K. (1962b). Food selection by sikworm larvae, swallowing factor. Nature, 194:754-755.

Hirao, T. and Ishikawa, M. (1964). J. Seric. Sci. Jap. 33:277.

Huber, R. and Hoppe, W. (1965) Die kristall-und molekulstrukturanalyse des Insektenver-puppungshormons Ecdyson mit der automatisierten Faltmplekulmethode. Chem. Ber., 98:2403-2424.

Imms, A. D. (1957). A General Text Book of entomology, Methuen and Co. Ltd., London.

Ito, T. and Horie, Y. (1966). Utilization of steroids and related compounds by the silkworm ***Bombys mori*** L, Annot. Zovl. Jap., 39:1-5.

Ishikawa, S. (1967). Maxillary chemoreceptors in the silkworm olfaction and taste II. Proceedings of the second International symposium, Tokyo, September, 1965.

Kawaguchi, E. (1926):Polyspermy in ***Bombyx mori*** L. Sangyo Sympo., 34:38-41.

Kagure, M. (1933). The influence of light and temperature on certain characters of the silkworm, ***Bombyx mori***. J. Dept. Agr. Kyushu Univ., 4:1-93.

Kikkawa, H. . (1943a). Origin of the egg pigment in ***Bombyx mori*** and the mechanism of its formation. Bull. Seric. Exp. Sta. Jap., 11:311-345.

Kikkawa, H. (1943b). Contribution to the knowledge of the enzymes working on the tryptophan metabolism in insects. Zool. Mag. Tokyo, 55:254-262.

Kobayashi, T. (1962). Number of gonial cells of the silkworm at early stages around hatching. Ann. Rept. Nat. Inst. Genet., Jap. 12:91-92.

Karlson, P. (1956). Biochemical studies on insect hormones. Vitam. Horn. 14:227-266.

Karlson, P. and Shaaya, E. (1964). The ecdyson titre of developing insects. I. A method of determing the ecdyson content. J. Insect Physiol. 10:797-804.

Krishnaswami, S., Narasimhanna, M. N. and Kumaraj, S. (1978). Sericulture Manual–2 Silkworm Rearing, F. A. O., Rome.

Legay, J. M. (1957). La prise de nourriture chez le Vers a soie (Doctoral Thesis, University of Paris, Paris, France).

Legay, J. M. (1958). Recent advances in silkworm nutrition. Ann. Rev. Ent. 3:75-86.

McCracken, I. (1990). Heriditary of race characters univoltinism and bivoltinism in the silkworm (***Bombyx mori***):A case of non-Mendelian inheritance. J. Exp. Zool., 7:747-764.

Morohoshi, S. (1950). Physiological genetics in ***Bombyx mori***. Meibundo, Tokyo.

Morohoshi, S., andMezaki, M (1957). Studies on the voltinism in the silkworm, ***Bombyx mori***. IV. On the locus of a major gene controlling voltinism. J. Seric. Sci. Jap. 26:257.

Magadum, S. B., Dutta, R. K., Remadevi, O. K., Mukesh Mohan and Choudhury, O. K. (1994). Evaluation of semi-synthetic diet for young age silkworm, ***Bombyx mori***. Indian J. Seric., 33(1):68-69.

Magadum, S. B., and Shivakumar, G. (1995). Formulation of semi-synthetic diet for young age silkworm, improvement of semi-synthetic diet and their bioassay. Annual Report, Central Sericultural Research and Training Institute, Mysore, India, pp. 106-107.

Nakamura, T. (1932). The inheritance of the moulting chacter in the silkworm. ***(Bombyx mori***, L). Rep. Seric. Exp. Sta. Chosen, 3:39-84.

Nagatomo, T. (1942). Inheritance of the voltinism in the silkworm, ***Bombyx mori***. J. Sci. Agr. Soc. Tokyo, 13:114-115.

Nakanishi, K., Koreeda, M., Sasaki, S. Cheng, M. L. and Hsu, H. Y. (1966). The structure of ponasterone A, an insect moulting hormone from the leaves of ***Podacarpus nakait*** Hay. J. Chem. Soc. Chem. Commun., 915-917.

Przibram, H. and Megusar, F. (1912). Wachstums-messungen an ***Sphodomantis bioculata*** Burn., 1. Lange und Masse. Arch. Entw. Mech. Org. 34:680-741.

Robert L. Patton (1963). Introductory Insect Physiology, W. B. Saunders Company, London.

Sado, T. (1965). Ann. Rep. Nat. Inst. Genet. Japan, 16:40.

Snodgress, R. E. (1935). Principals of Insect Physiology, McGraw-Hill Book Company, INC.

Tazima, Y. (1964). The genetics of Silkworm. Logos Press Ltd., London.

Tazima, Y. (1978). The Silkworm:an laboratory tool. Kodansha, Ltd., Tokyo, Japan.

The physiology of Insecta, (Ed. by Morris Rockstein), 1974, Academic Press, New York, London.

Torii, K. and Morii, K. (1948) Rept. Inst. Sericult. Sci., 2:3-11.

Toyama, K. (1906a). Studies on the hybridology of insects. I. On some silkworm crosses with special reference to Mendel's law of heredity. Bull. Coll. Agr. Tokyo Univ., 7:259-393.

Toyama, K. (1906b). Mendel's law of heredity as applied to the silkworm crosses. Biol. Zbl., 26:321-334.

Toyama, K. (1912a). On the varying dominance of certain white breeds of the silkworm, ***Bombyx mori*** L. Z. I. A. V., 7:253-288.

Toyama, K. (1913). Maternal inheritance of Mendelism (Frist Contribution). J. Genet., 2:351-404.

Training Mannual on silkworm physiology, Zhejiang University, P. R. China. (1994).

Text book of Tropical Sericulture(1975). Japan Overseas Corporation Volunteer 4-2-24 Hiroo, SIBUYA-KU, TOKYO, JAPAN.

Umeya, Y. (1925). On the experiments of ovarian transplantation and blood transfusion in silkworm, with special reference to the alteration of voltinism. Jap. J. Genet., 3:155-182.

Umeya, Y. (1926b). Experiments of ovarian translation and blood transfusion in silkworms with special reference to the alteration of voltinism. Bull. Seric. Exp. Sta. Chosen, 1:1-26.

Waldbauer, G. P. (1968) The composition and utilization of food by Insects. Adv. Insect Physiol., 5:229-288.

Watanabe, K. (1918b). Studies on the voltinism in the silkworm, ***Bombyx mori*** I. Inheritance of bivoltine vs. quadrivoltine. Bull. Seric. Expt. Sta. Jap., 3:439-456.

Watanabe, K. (1919). Studies on the voltinism in the silkworm, ***Bombyx mori*** II. Inheritance of Univoltine vs. multivoltine. Ibid. 4:87-106.

Watanabe, K. (1924). Studies on the voltinism in the silkworm, ***Bombyx mori***. Ibid. 6:411-415.

Watanabe, T. (1958). Substances in mulberry leaves which attract silkworm larvae. Nature, 182:325-326.

Williams, C. M. (1956). The juvenile hormone of insects. Nature, 178:212-213.

Wigglesworth, V. B. (1950). Principles of Insect Physiology. Methuen and Co., Ltd., London, England.

Index